TECHNOLOGY AND RESOURCE USE
IN MEDIEVAL EUROPE:
CATHEDRALS, MILLS, AND MINES

TECHNOLOGY AND RESOURCE USE
IN MEDIEVAL EUROPE:
CATHEDRALS, MILLS, AND MINES

edited by

Elizabeth Bradford Smith and Michael Wolfe

Ashgate

Aldershot • Brookfield USA • Singapore • Sydney

British Library Cataloguing-in-Publication Data

Technology and Resource Use in Medieval Europe: Cathedrals, Mills, and Mines
1. Middle Ages – Congresses 2. Technology – History – Congresses
3. Science, Medieval – Congresses
I. Smith, Elizabeth Bradford II. Wolfe, Michael
609'

U.S. Library of Congress Cataloguing-in-Publication Data

Technology and Resource Use in Medieval Europe: Cathedrals, Mills, and Mines /
Edited by Elizabeth Bradford Smith and Michael Wolfe
p. cm. Essays from a conference held in April 1995. Includes index.
1. Technology – History – Congresses. 2. Civilization, Medieval – Congresses.
I. Smith, Elizabeth Bradford II. Wolfe, Michael
T17.C38 1997   97-39901
609.4'09'02 – dc21   CIP

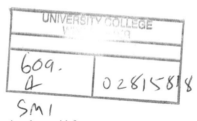

Published by   Ashgate Publishing Limited
                          Gower House, Croft Road,
                          Aldershot, Hampshire GU11 3HR
                          Great Britain

                          Ashgate Publishing Limited
                          Old Post Road
                          Brookfield, Vermont 05036
                          USA

ISBN 0 86078 670 6

Printed in Great Britain by Galliard (Printers) Ltd, Great Yarmouth

Contents

List of Contributors

Patrice Beck, École des Hautes Études, Paris
Niall Brady, Trinity University, Hartford CT
Philippe Braunstein, École des Hautes Études, Paris
Lynn Courtenay, University of Wisconsin
David Crossley, University of Sheffield
Bert Hall, University of Toronto
Richard Holt, University of Birmingham
Kristina Luce, Miami University, Ohio
Robert Mark, Princeton University
Michel Philippe, École des Hautes Études, Paris
Sergio Sanabria, Miami University, Ohio
Elizabeth Bradford Smith, Pennsylvania State University
Paolo Squatriti, University of Michigan
Michael Toch, Hebrew University of Jerusalem
Michael Wolfe, Pennsylvania State University-Altoona

List of Illustrations

Introduction: New Perspectives on Medieval Technology and Resource Use

Elizabeth Bradford Smith and Michael Wolfe

The essays in this collection are the fruit of a conference devoted to the study of medieval technology sponsored by the Center for Medieval Studies at the Pennsylvania State University in April 1995. Taken together, they once again dispel the regrettable public misperception that medieval people somehow had to toil in a world bereft of technical innovation and ingenuity. Nothing, of course, could be further from the truth, as these specialists in the field all amply demonstrate. Yet our authors do more than just redeem the Middle Ages as a time of considerable technological development; they also show the manifold ways in which the technologies of building construction, manufacture, and metallurgy shaped and were in turn shaped by broader forces of culture, social identity, political ambition, and the local environment.

The collection opens with two essays devoted to that most celebrated example of medieval technological expertise, the Gothic cathedral. Another two then examine lesser known, but equally revealing vernacular structures, such as hospitals and barns. Robert Mark has been studying the engineering of large-scale medieval structures, especially cathedrals, over the past twenty-five years. His essay begins by presenting a synthesis of his work as it concerns design strategies used by Gothic masons. Mark reinforces here what he has shown in previous publications—that in addition to pragmatic conditions, such as available materials and economic factors, medieval masons relied primarily on experimentation and observation, rather than on any mathematical or geometric theory. He then goes on to look at the situation in the Renaissance and Baroque eras, where treatises of design based on the rules of classical Greco-Roman architecture prevailed. These learned books tempted many an architect to abandon the medieval reliance on experimentation at the work site in favor of a more distant relation to the building. As a result, structural innovation was no longer a primary feature of most buildings. Nevertheless, some unprecedented structures—such as Brunelleschi's dome in Florence, the cupola of St Peter's in Rome, and Wren's dome for St Paul's in London—were erected in this period. These, Mark points out, could never have been built by simply following the rules.

The architects responsible for these structures had to rely on knowledge of traditional building practice and on observation. Thus, Christopher Wren's great dome combines Renaissance design with the structural tradition of High Gothic architecture. As we will see, this will hardly be the last example of the surprising continuity of medieval technological traditions up to 1700.

Sergio Sanabria's ongoing meticulous survey of the cathedral of Metz has thus far yielded valuable results concerning the structure of one of Europe's largest Gothic churches. Hitherto marginalized by its position on the periphery of the French-speaking world and, more importantly, by having long been mistakenly dated to the latter half of the thirteenth century, Metz in fact was directly linked to mainstream influences from Reims. Moreover, as Sanabria demonstrates, rather than being merely derivative, Metz often displays innovative and daring solutions to structural problems associated with large-scale buildings of the Gothic era. In the present essay, Sanabria analyzes the slender and delicate upper buttresses of the nave at Metz to determine how closely they approach viable structural limits. His conclusions, that they are loaded within 20 per cent of their tension limit, as compared to the 30-35 per cent common in modern engineered structures, shows that the masons of Metz were aware of the strengths and limitations of the medium in which they worked. In sum, the buttresses of Metz now stand as yet another proof of the sound and elegant building practices which emerged from the empirical methods of the master masons of the Middle Ages.

The contribution by Lynn Courtenay on large-scale timber work of the High Middle Ages combines a short introduction to medieval timber construction practices with a close look at major examples of three different types of building and of carpentry assembly from the thirteenth century. The buildings discussed—the Knights Templars' Wheat Barn at Cressing Temple Manor, Essex; the nave roof of the cathedral of Notre-Dame of Paris; and the main ward of the Byloke Hospital in Ghent—all cover vast spans with roofs of monumental grandeur. Especially illuminated in this study is the way in which each of these structures marries efficiency of design with a thoughtful and economical use of timber. Indeed, analysis of these buildings reveals the medieval carpenter of northern Europe as an integral part of an interrelated system which included building technology, woodworking processes, prudent management of woodland resources, and overall cost control. Courtenay convincingly demonstrates how the master carpenters produced the elegant timber work we admire so much today without depleting available woodlands only by carefully maintaining a sustainable balance between the reciprocal needs of each of the components in this system.

Buildings not only encapsulate an intricate process of design and construction, but also often serve as potent symbols of cultural and social authority, Niall Brady reminds us in his detailed look at Gothic barns in medieval England. A congeries of factors shaped the forms and functions of this long-neglected type of medieval vernacular architecture. Building a barn represented a considerable capital investment on the part of the lord, in some cases up to a year's revenues for a manor. The economics of barn-building clearly offers us another example of the way in which rational calculation, particularly the lord's efforts to minimize risk and maximize profit, shaped agrarian England. Yet, as Brady astutely notes, the pursuit of profit also drew on the prevailing notions of prestige and status associated with manorial lordship. The addition of architectural ornamentation to a thirteenth-century barn, for example, underscored the symbolic significance of a building which, perhaps more than any other structure in the countryside, expressed the lord's authority over his tenants. The desire to overawe therefore explains the otherwise curious fact that many barns possessed considerably more storage capacity than was reasonably needed by a manor. Medieval barns thus both contained and displayed the lord's wealth to those whose labor he exploited. The technologies employed to construct buildings during the Middle Ages established through architecture powerful modes of communication intended to reinforce the traditional social hierarchy. Incidentally, this also explains why the violent peasant revolts later in the fourteenth century, especially Watt's Rebellion of 1381, targeted such buildings in their attempt to overthrow the oppressive manorial regime under which they toiled.

The essays in part two shift attention to the ways in which medieval people tried to harness the power of water and wind. As David Crossley shows, water mill technology in the British Isles held an important place in the medieval economy both for agricultural as well as metallurgical production. Crossley's piece offer us an update of ongoing archaeological investigations into use of water power in pre-industrial England. His site-specific findings complement Squatriti's more generalized, interpretative approach to the topic of medieval hydraulic technology. Despite different methodologies, both archaeologists and historians share a common appreciation of the central importance of local topography in explaining the form and function of particular water mills. Crossley illustrates the often subtle factors behind the siting of a mill in relation to its water source and its application. Careful attention is also devoted to the construction materials and techniques which mill owners used in their quest to maximize profits and lower costs. The continuity and adaptability of long-established water mill designs again, as in Italy, ruled the day. A particularly compelling aspect of Crossley's essay is the meticulous manner in which he reconstructs out of remaining physical evidence the system of sluices, catchments, and races that had to be precisely coordinated—sometimes over distances of several miles—to enable the mill to run efficiently. Although sources reveal the existence of horizontal-wheeled mills in Anglo-Saxon England, there developed in the wake of the Norman Conquest a decided preference for vertical-wheeled mills across the country. In no time, it seemed, thousands of water mills churned daily to grind corn, process ore, or full cloth. Indeed, as archaeological evidence makes clear, medieval water mill technology flourished in England well into the eighteenth century, until it was displaced by steam power in the early Industrial Revolution. As Crossley suggests, the next step for archaeologists and historians will be to link from the local to national levels the role played by water mills in shaping the economic and social development of preindustrial England.

Paolo Squatriti substantially revises our understanding of the place of water mill technology in early medieval Italy. Marc Bloch's contention that water mills represented a distinctly medieval response to demographic and economic conditions on manorial estates—a view long reiterated by historians—fails to hold up in light of recent archaeological findings from the ancient Roman world. Indeed, from the sixth century onward, water mills of all sorts formed an integral, long-established part of the rural economy adapted to the features of local terrain and society up and down the Apennine peninsula. The sudden appearance of numerous references to water mills in eighth-century cartularies did not, as Bloch believed, represent the advent of this technology, but rather reflected the higher survival rate of this type of document, according to Squatriti. This abiding continuity in hydraulic technology between the classical and early medieval periods forces us to revise the notion that water mills developed in early medieval Europe in response to labor shortages resulting from the gradual disappearance of slavery and supposed demographic contraction after C.E. 500. The fact that water mills existed in Antiquity alongside slave-based *latifundia* weakens this correlation, as does recent evidence suggesting much lower levels of population decline in early medieval Italy. Far from reflecting more straitened economic circumstances and a desire to save labor, the heightened use of water mills during the eighth and ninth centuries actually responded to the rising productive vitality of manorial estates and gender shifts in the organization of work. Although Bloch's contention that water mills offered profit-hungry lords a means to extract surplus wealth from producers holds up in some instances, new evidence shows that many water mills were often built and run by urban entrepreneurs and peasants. The enthusiasm for water mills among these diverse groups thus arose from a variety of complex factors related to contemporary notions of utility, prestige, and taste. The advent and conquest of water mills in early medieval Italy was, in short, as much a cultural as an economic phenomenon.

Richard Holt's essay offers a comprehensive overview of the impact of mechanization on the medieval English economy. As in early medieval Italy, water mills formed a familiar part of the English landscape, widely distributed across the shires, according to the late eleventh-century Domesday Survey. Material constraints limited the size and power of most mills, Holt observes, since they used primarily timber, not iron, down to the eighteenth century. Nevertheless, water mills generated a fairly reliable profit for landlords, though access to milling technology often existed outside of seigneurial control as seen in the ubiquitous presence of household querns, private horse mills, and—in the thirteenth century—windmills. Except along the blustery coast of East Anglia, the use of windmills paled in comparison to water mills (a point, again, in contrast to the views of both Marc Bloch and Lynn White) largely due to conservative estimates of their profitability. Holt estimates that the number of water mills doubled from the eleventh to the fourteenth centuries in response to rising population and reflective of the central place of bread and ale in the medieval diet. Even so, he suspects that a large portion of the grain used for bread and ale was processed at home using handmills. Milling technology also found a number of manufacturing applications, particularly in textiles, though Holt believes that profit margins narrowed in such ventures. For the most part, such instances of mechanization in preindustrial England through the application of water and wind water remained insignificant until the very end of the Middle Ages. For Holt, the key to understanding the operation of the medieval English economy lay not in technology (a modern conceit at best), but in labor organization and artisanal traditions that were, in his estimation, much better attuned to the needs of household-based, urban, craft production. How all this began to change in the sixteenth century—protoindustrialization—up to the revolutionary mechanization of the eighteenth-century Industrial Revolution—awaits further research, however.

Michael Toch continues this re-evaluation of the relationship between labor and technology by examining the agrarian economy of medieval Germany. He questions Lynn White's original contention, offered in his *Medieval Technology and Social Change* (1962), that the development of better means of harnessing horse power, more sturdy plows, and the three-field system had revolutionized medieval agriculture by the eleventh century. For Toch, the 'agricultural revolution'—if we insist on speaking in such terms—differed markedly from the one envisaged by White, whose equation of technology with human progress reflects the optimism of a generation ago. In fact, agricultural output rose most dramatically over the twelfth and thirteenth centuries, well after the advent of these supposedly 'revolutionizing' technologies. The privileging of technology over social and cultural considerations, evident not only in modern Western society but in post-colonial societies as well, forces us to examine the relationship between technology, society, and culture during the Middle Ages. Toch attributes the vitality of German agriculture during the Middle Ages to new organizational and cultural adaptations that intensified human labor, not labor-saving devices. Improved tools, when they appeared, coexisted rather than displaced older tools, as was the case with the scythe and sickle. Such technological heterogeneity continued into the early modern era. Technological changes sought to maximize the use of natural resources rather than save human labor until the population declines occasioned by the Black Death put a premium on supplementing human hands. The rise of new tenancy arrangements that granted more autonomy to peasants spurred the development of a more commercialized agricultural economy, too. Changing modes of production, ranging from arable husbandry to grain cultivation as well as specialized crop production, also brought profound changes that did not require the advent of new technologies. The availability of adequate labor and capital, not technology, proved to be the decisive variables in shaping the direction of German agricultural development over the Middle Ages and early modern period. In the end, social and cultural changes, not technology, accounted for ongoing and—cumulatively speaking at least—dramatic growth of the German medieval agrarian economy.

The interplay of natural resources, technology, and human needs is explored at the micro-level further in the essay by Beck, Braunstein, and Philippe on wood, iron, and water in the Othe Forest at the end of the Middle Ages. This collaborative essay joins together an impressive array of approaches, some archival and others archaeological, to trace the evolution of iron mining and foundry operations in the case of the Othe forest, located on the border between Champagne and Burgundy. Such a concentrated regional focus enables us to observe the different, complex ways in which terrain, water and forest resources, settlement patterns, and the shifting demands of market forces all combined to shape the rise and ultimate demise of mining sites in this region during the Middle Ages. Painstaking topographical surveys have located a considerable range of different sized sites, while manorial account books and legal records provide clues about the impact of iron mining on the local economy and society. Competition, from both the iron market and other exploitable forest resources, not exhaustion of local iron mines as was once commonly supposed, account for the slow decline of mining in the Othe after 1400. Additionally, property owners increasingly invested their money in water mills, which could be geared for a wider variety of tasks than could foundries, which existed only to smelt iron. This change reflected property owners' adoption of sophisticated business strategies responsive to sensitive market and environmental factors. Braunstein's team proposes that future investigation will seek to discern more clearly the variegated nature of local iron production in different, distinctive areas within the region. At the same time, reconstructing the production history of other forest industries will broaden our understanding of the place of iron mining in the wider economy.

Among iron's many uses in late medieval Europe was the manufacture of gunpowder weaponry—a topic explored in Bert Hall's compelling social history of early hand-held firearms. Such weapons, he argues, developed in late medieval southern German cities for reasons that had more to do with urban politics and society than military matters. Municipal self-defense, directed against both external threats and potential internal disorders, fell to citizen militias mainly composed of freemen from a city's guilds. The effectiveness of such amateurs, who—as the records show—usually despised the tedious business of mustering and patrolling, was substantially increased by outfitting them with 'user-friendly' weapons, such as first the crossbow and then firearms. Hall therefore casts doubt on the long-standing contention that firearms revolutionized war making at the end of the Middle Ages by discerning the important continuities between the crossbow and gunpowder weaponry in terms both of manufacture and use. Similarly, market forces and guild traditions also powerfully shaped the early development of the arms industry, especially in the areas of technological innovations and the dispersed nature of actual production. Another important factor that contributed to the triumph of this new weaponry was the ongoing reduction of manufacturing costs brought about over the fifteenth century by improved techniques of producing both weapons and gunpowder. In these ways, the skills associated with firearms actually lay in their production, not use. The new age of gunpowder weaponry thus took the forms it did in large part due to the strong craft traditions and culture of late medieval cities.

These essays offer a variety of perspectives by which to approach a better understanding of the place of technology in medieval European society. All of them show that discrete technologies in the areas of building construction, manufacturing, and metallurgy actually offer us highly revealing points where human and environmental factors intersected and interact with each other. In this sense, the place of technology in medieval society must be seen in terms of the complex, interconnected systems of resource use and management, transportation, labor organization and expertise, and modes of distribution and consumption that are every bit as intricate and pervasive as the ones we so routinely trumpet in our modern, high-tech world.

Part I

Stone and Wood

1

Technological Innovation in High Gothic Architecture

Robert Mark

INTRODUCTION

In the second half of the twelfth century, master builders of the Île-de-France began constructing the great vaulted spaces enclosed by large expanses of glass and supported by supple, skeletal stone structure that we identify as High Gothic.[1] Mercantile and ecclesiastical centers vied with one another to attain extreme building height in church construction—not unlike the competition between American cities in lofty steel-framed office buildings some seven centuries later.

Apart from their taller but usually far more heavily constructed towers, the height of the lofty, main vaulted vessels of High Gothic churches is striking. The keystones of the high vaults of Beauvais Cathedral (begun in 1225) rise a full 48 m above the floor (a cross-section is shown in fig. 1.1); and a high-peaked timber roof supported on parapet walls raised above the level of the vault adds another 20 m to the elevation. As Gothic buildings began to reach up to these heights, wind forces became a far more important factor in design. Not only did the taller structures present larger lateral surfaces to the wind (i.e., greater 'sail area'), but they were also subject to appreciably higher wind speeds at the higher elevations. This is particularly critical since wind pressure, and suction on a building's leeward side, are proportional to the square of the wind speed.[2] Tall building structures were then called upon to resist substantial lateral wind forces as well as dead weight. Problems of building were exacerbated also by what must have seemed to be the insuperable costs of quarrying and transporting vast amounts of stone to the building site, often from great distances, and of shaping and setting it into place. Minimizing quantities of stone used for structure could well reduce this expense; and it also reduced foundation loadings and associated settlements.

---

[1] See Jean Bony, *French Gothic Architecture of the 12th and 13th Centuries*, Berkeley, 1983, pp. 26-43.

[2] R. Mark, *Experiments in Gothic Structure*, Cambridge, 1982, pp. 22-4.

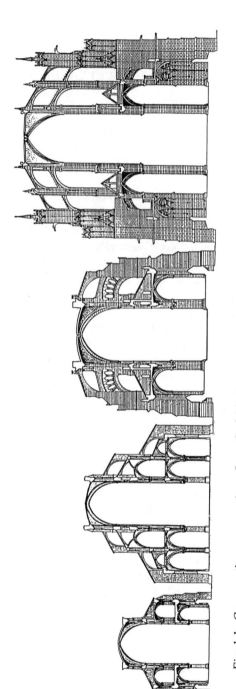

Fig. 1.1. Comparative cross-sections: Laon Cathedral, nave, ca. 1175; Notre-Dame-de-Paris, nave, ca. 1180; Chartres Cathedral, nave, begun 1194; Beauvais Cathedral, choir, begun 1225.

All of these factors seem to have led, in the late-twelfth century, to the introduction of the exposed flying buttress that allowed a great reduction in the mass of building walls and redefined the style of Gothic churches.[3] The process of design that led to such splendid achievement in a pre-scientific era will be explored and then compared to the later, more 'modern' approach to architectural design that displaced it.

HIGH GOTHIC DESIGN
Unfortunately, the few surviving documents that have come down to us from the High Gothic era provide precious little information on design strategy. The most important of these is the early thirteenth-century sketchbook of Villard de Honnecourt that contains partial plans and views of several contemporary buildings along with a number of drawings and some slight text dealing with such topics as the setting out and cutting of stone, timber roof framing, rudimentary machines (such as a screw jack), and applied geometry.[4] Villard indicates no rules for design other than inscribing geometric figures—circles, squares, triangles, and octagons—to configure architectural elements. No canons are offered, geometric or otherwise, that might govern sound structure. Yet, the efficacy of High Gothic structure reveals that structural lessons learned from previous building projects must have contributed to the success of new designs. The constructional organization of that time that allowed apprentices to rise through the ranks, and even to assume the position of master builder-designer, insured familiarity with earlier buildings. In effect, these served as approximate models for the new, taller buildings; but the earlier buildings could not by themselves have pointed to structural solutions for coping with the forces of high winds.

Since the sequence of construction of Gothic churches usually called for erecting the upper portion of the main vessel one bay at a time, I have suggested that the first completed bays could have been used to test new structural ideas and thus to fix the form of succeeding bays. This experimentation was largely accomplished, I believe, by observing cracks developing in the tension-sensitive lime mortar used to cement the cut stones, and then by modifying the form of the structure in order to expunge the cracking.[5] The intrinsic power of this approach is underlined in the recent commentary of engineer Henry Petrosky concerning contemporary design: 'it has long been practically a truism among practicing engineers and designers

[3] R. Mark and W.W. Clark, 'Gothic Structural Experimentation', *Scientific American*, 251 (November 1984) 176-85.
[4] T. Bowie, ed., *The Sketchbook of Villard de Honnecourt*, Bloomington, 1959.
[5] Mark, *Experiments*, p. 119.

that we learn much more from failures than from successes. Indeed, the history of engineering is full of examples of dramatic failures that were once considered confident extrapolations of successful designs; it was the failures that ultimately revealed the latent flaws in design logic that were initially masked by large factors of safety.... Design studies that concentrate only on how successful designs are produced can thus miss some fundamental aspect of the design process'.[6]

The reliance on observation of failure by the builders is evident also from an examination of the modes of High Gothic vaulting. With few exceptions, ribbed sexpartite vaulting (six-part square-planned ribbed vaults covering two bays) was specified for the high bays of all the major Gothic churches of the second half of the twelfth century, including the giant cathedrals of Notre-Dame-de-Paris and St Étienne de Bourges. Then quite suddenly, at the beginning of the thirteenth century, all High Gothic churches, including Chartres Cathedral, deployed quadripartite high vaults. While a stylistic explanation has generally been offered for the transition, far more compelling grounds are provided by examining the forces involved in supporting the high vaults during their construction. Since vault erection was necessarily preceded by that of supporting piers, walls, and flying buttresses, the outward, horizontal thrusts of a vault after the removal of the temporary, timber centering on which it was assembled (so the centering could be reused in adjacent bays) is resisted by the same sturdy structural elements as in the finished church. On the other hand, longitudinal thrusts, lengthwise along the axis of the church, would need to be stabilized by temporary timber shoring (as illustrated in fig. 1.2) until the erection of the adjacent bay of vaulting that acts as a longitudinal brace in the finished building. Numerical-computer modeling of the sexpartite vaults of Bourges Cathedral (the largest building having the older type of vaulting) indicated the longitudinal thrust to be 19,000 kg. For the roughly equivalent scale, quadripartite vaulting of Chartres Cathedral, longitudinal thrust is only about half as much (fig. 1.3).[7]

The constructional problems engendered by the intensity of these longitudinal thrusts do not appear to have been crucial in the early Gothic churches where the vault springing was anchored into massive walls stiffened by a vaulted gallery. The countering of this thrust only became acute with the demand for greater clerestory height and accompanying larger fenestration which brought an end to the practice of springing the vault from a solid wall. Lacking our scientific methods of analysis to determine the vault thrusts, the

---

[6] H. Petrosky, *Design Paradigms*, New York, 1994, pp. 1, 10.

[7] W. Taylor and R. Mark, 'The Technology of Transition: Sexpartite to Quadripartite Vaulting in High Gothic Architecture', *Art Bulletin* 64 (December 1982) 578-87.

Fig.1.2. Longitudinal shoring used in construction of high vaults.

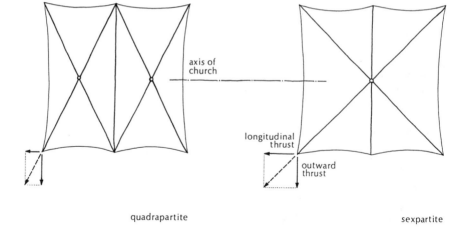

quadrapartite   sexpartite

Fig. 1.3.   Comparative plans of quadripartite and sexpartite vaulting showing components of vault reactions.

early Gothic builder surely must have encountered cracking in the tall supporting piers in the course of construction, and this led to a redesign of the high vaults. As with the flying buttresses, quadripartite vaulting was adopted primarily for reasons of structure: flying buttresses provided dependable support for the transverse thrusts of the high vaults while quadripartite vaulting reduced the longitudinal thrusts to manageable levels during construction.

LATE GOTHIC DESIGN
After 1280 there was a notable decline in northern-French tall-building activity. Although this has often been ascribed to a collapse, due to a hidden design flaw, of the original (quadripartite) high vaulting of Beauvais Cathedral that occurred in 1284, it was mainly due to a significant downturn in the local economy, and soon thereafter also by the outbreak of the Hundred Years War between England and France.[8] Even so, enormous Late Gothic churches constructed in Metz, Narbonne, Palma (on the island of Majorca), Milan, Seville, and Ulm attest to the continuing vitality of the Gothic style throughout the rest of Europe; although the southern churches, generally topped by far less steep roofs than those of the north and subjected to gentler

[8] On the cause of the collapse at Beauvais in 1284, see Mark, *Experiments*, pp. 58-75. On economic history, see J. Strayer, *On the Medieval Origins of the Modern State*, Princeton, 1970, pp. 57-8.

winds, did not present such critical problems of wind loading.[9] The written record of a meeting held in 1401 between French, German, and Italian specialists to confer on the ongoing construction of Milan Cathedral allows another rare glimpse into the late medieval design process. The so-called *Milan Expertise*, that seems to deal mainly with geometric rather than structural arguments, however, suffers from being a summary of the discussions by an uncomprehending scribe.[10]

A wider window on design technique opens near the end of the fourteenth century with the preservation of several architects' notebooks.[11] Although presented in more detail than in the sketchbook of Villard de Honnecourt, the approaches used for shaping building elements are based on similar geometric schemes; and these architects, too, do not deal directly with questions of structure. Specific design rules for structure—based on observations of comparable extant buildings—are found in the *Instructions* written in 1516, apparently for the benefit of his son, by Lorenz Lechler, a master mason at Heidelberg.[12] In addition to presenting geometric schemes for spatial planning, Lechler gives specific advice on structural details such as wall thickness, window-opening sizes, and buttress and vault-rib dimensions. While Gothic architecture still flourished in Germany after the fourteenth century, a modified type of basilica-church interior was more commonly used. In these late-Gothic so-called hall churches, all three aisles were made almost equal in height. By eliminating the high, central clerestory of the High Gothic nave, a large floor plan could be realized at far less expense. Flying buttresses were replaced by wall buttresses—or simply, reinforcing the relatively low exterior walls by projecting masonry 'legs' between the windows. Nonetheless, a great timber-framed roof covering all three aisles (as illustrated in fig. 1.4) would still exhibit a tall exterior.

Although Lechler does not clearly state their interdependence, the dimensions of all the building elements are proportional to the interior span of the central aisle which he advocates keeping between 6 and 9 m. For example, the recommended height of the central-aisle vault keystone is given as one and

---

[9] Mark, *Experiments*, p. 98.

[10] John White, *Art and Architecture in Italy 1250-1400*, 2nd ed., p. 521, characterizes the *Expertise* as 'over-simplified notarial summaries of long and complex arguments...written in very crude and occasionally uncomprehending Latin'. See also J.S. Ackerman, '"Ars sine scientia nihil est": Gothic Theory of Architecture at the Cathedral of Milan', *Art Bulletin* 31 no. 2 (June 1949) 83-111.

[11] See, L.R. Shelby, *Gothic Design Techniques: The Fifteenth-Century Design Booklets of Mathes Roriczer and Hanns Schmuttermayer*, Carbondale, 1977.

[12] L.R. Shelby and R. Mark, 'Late Gothic Structural Design in the "Instructions" of Lorenz Lechler', *Architectura* 9 (1979) 113-31.

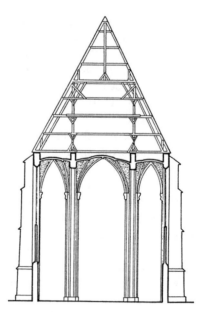

Fig. 1.4. Cross-section through a typical late Gothic hall church.

one-half or two times the span of the central aisle. Wall thickness is essentially one-tenth the span, as is the width of the outstanding leg of the wall buttress at ground level. The buttress extends about two-tenths of the span from the outside edge of the wall, giving a total (wall plus buttress) depth at ground level of three-tenths of the span of the central aisle. Thus, for a church with a 9 m central-aisle span, the vault keystone elevation may be 13.5 or 18 m, while the total depth of buttressing in both cases is about 2.7 m. Holding the buttress depth constant for a range of building heights makes sense only with regard to resisting the outward thrust of the vault where both the vault thrust and the stability of the buttress against overturning increase as a function of the cube of the building scale. Lechler's rule does not account for the effect of wind on the tall roofs, but since his buildings were intended to be relatively small, the omission may not have been serious for local construction. Nonetheless, the turning away from observation of failure and the adoption of the more 'modern' written design rules must have stifled structural innovation. Indeed, the structural feats of the High Gothic were not to be repeated.

CLASSICAL DESIGN IN THE RENAISSANCE
Bear in mind that although Lechler, working in Germany, continued to be attentive to Gothic design, he was doing so when traditional building practice

in Renaissance Italy no longer served as the wellspring of architectural knowledge. It had been supplanted by a reliance on the authority of ancient architecture derived from the observation of Roman ruins and, following its 'discovery' in 1415, the Augustan treatise of Vitruvius which served also as a model for Renaissance architects to follow in their own writings. The career of Leon Battista Alberti (ca. 1404-72) epitomizes this new phenomenon: a humanist by training who examined Roman ruins while working as secretary at the Vatican, he later employed craft-architects on site who directed the construction of his designs. Alberti also wrote a ten-volume treatise (following the Vitruvian model) on the art of building. Published in the mid-fifteenth century, *De re aedificatoria* is the first known treatise since Vitruvius to be entirely devoted to architecture.[13] It was then followed during the succeeding century and a half by a flurry of Italian architectural treatises; and facilitated by the development of printing, translations of these became accessible throughout Europe by the end of the sixteenth century. An important consequence was that structural innovation became less of an issue in architectural design. This may be perceived even in the general treatment of walls. Unlike the almost transparent northern medieval wall, walls of archetypal Renaissance buildings were built in imitation of the massive works of Imperial Rome. Niches and applied pilasters were intended to impart a similar monumental quality.[14]

But because of the unusually large scale of their lofty central domes, the great domed cathedrals of the Renaissance which include the cathedrals of Florence and of London, and the new basilica of Rome, could never have been realized by simply following rules. The first of these, at Florence, called for an immense dome over earlier Gothic construction. The octagonal opening of the supporting drum to be covered by the dome was 42 m across the sides of the dome base, and about 55 m above ground level—a scale that all but ruled out the possibility of erecting the dome on timber centering supported from the

---

[13] Leon Battista Alberti, *On the Art of Building in Ten Books*, J. Rykwert, N. Leach, and R. Tavenor trans., Cambridge, 1988, ix. To Alberti's credit, despite his immersion in Latin Humanism, to reach a wider audience he employed the vernacular in his own writing. In so doing, he is credited as being the first major writer to help reestablish Italian as the literary language of Italy. See G.C. Sellery, *The Renaissance: Its Nature and Origins* Madison, 1950, p. 110.

[14] Nor would technological innovation seem to have been encouraged by Renaissance culture. As Sellery recounts: 'The Humanists repeatedly make the point that they do not seek to be innovators, but rather restorers of that which had already existed in the Golden Age of the past. How very different the view of the natural scientists. They regard themselves as radical innovators, challenging the authority of the past, and they located their "Golden Age" in the future'. *The Renaissance*, p. 255.

floor below. Filippo Brunelleschi, the principal architect of the dome project that began in 1420, overcame the formidable problem of how to support the rising dome by constructing a pointed, intricately-ribbed double-shell brick-and-stone dome of enormous total thickness (about 4 m at its base) in horizontal full-ring layers to maintain stability.[15] Timber centering was then used to erect the fabric of the more horizontal upper portions of the dome as the spans from side to side are appreciably less than across the supporting drum below. The final stage of construction, the placement of a heavy stone central lantern was completed in 1462, sixteen years after the death of the designer. Brunelleschi's dome was considered the most magnificent technological and artistic feat of the age; but it did not escape the structural distress illustrated in fig. 1.5, despite his specification of a great timber chain and iron-clamped sandstone blocks encircling the dome in an attempt to resist outward thrusts.[16] The same type of distress would be evident also in the similarly-scaled dome of St Peter's Basilica in Rome begun almost a half-century later.

Under the direction of Donato Bramante, the first principal architect of St Peter's, construction commenced on the four massive piers that were to support the great dome. Bramante seems to have made no detailed design for the dome itself, but his piers set its diameter. It was left for Michelangelo to plan and to begin construction of the dome between the years 1546 and 1564. Following two decades of inactivity, the design was again revised by Giacomo della Porta, with the assistance of Domenico Fontana, who completed it during 1588-93. As at Florence, St Peter's dome incorporates a ribbed, thick, double shell of brick and stone as well as a massive stone lantern placed above its summit.

Even though their profiles are pointed (that of Florence more so than St Peter's) which serves to reduce their thrusts somewhat, both domes are plagued with problems of cracking due to their not being sufficiently buttressed. This becomes evident when we compare the cross-section through the dome of St Peter's with that of the ancient Pantheon whose dome has practically the same span (fig. 1.6). In the Pantheon, the dome's outer profile is relatively flat, and buttressing against the outward thrusts of the dome is well provided for by a massive concrete wall. At St Peter's, there is only the

---

[15] See Howard Saalman, *Filippo Brunelleschi: The Cupola of Santa Maria del Fiore*, London, 1980, pp. 112-34.

[16] Ongoing problems of cracking in the Florence dome were the subject of an article in the *New York Times* entitled 'Cracks in a Great Dome in Florence May Point to Impending Disaster', (28 July 1987: C3).

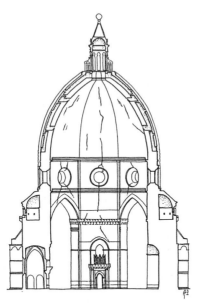

Fig. 1.5. Florence Cathedral: cross-section through the great dome showing typical tension crack patterns.

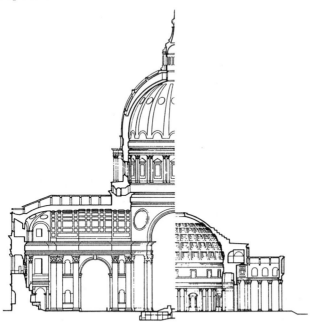

Fig. 1.6. Comparative cross-sections: St Peter's Basilica and Pantheon (Rome)

vertical portion of the relatively thin cylindrical drum below the dome to provide what has proved to be inadequate resistance. Della Porta placed two iron chains around the base of the dome to help prevent its spreading; but the problem of its spreading, and subsequent cracking, is exacerbated by the great weight of the dome and lantern which generates extremely large outward forces. Almost from the beginning the dome began to crack along meridians as it spread outward, and five more iron chains were added by the middle of the eighteenth century.[17]

Although the diameter of Christopher Wren's dome for the new St Paul's Cathedral in London (to replace the medieval cathedral destroyed by fire) was to be but three-quarters that of St Peter's, Wren seems to have been alarmed by reports of the cracking in the Roman Basilica. In typical Renaissance fashion, the walls of the cathedral are excessively heavy, but Wren realized that the design of the dome necessitated special care. There are several references to experimental dome models in his notes of 1694, and some later dome schemes dated as late as 1703 are not yet his last version.[18] The final design, constructed between 1705-08, is based on a majestic, light outer dome profile of lead-sheathed timber supported by an unseen chain-girdled brick cone, only 46 cm thick and which also supports a stone lantern of some 850 tons, and a separate brick dome, also 46 cm thick (fig. 1.7). The brick cone of St Paul's, formed by straight-line generators, is directly loaded by the heavy central lantern. Hence the cone, which also provides support for the outer, timber dome, experiences almost uniform compression rather than the pernicious tension endured by the more spherically-shaped domes of Florence and St Peter's. In striking contrast to St Peter's, Wren's single iron chain has proved sufficient to maintain the integrity of the relatively light-weight structure against outward thrusts. Indeed, the triumph of Wren's structural solution is manifest in its having been the model for all large-scale dome projects that followed St Paul's well into the nineteenth century.

Wren's vital contribution to large dome design was the hidden chain-reinforced structural cone; the essential idea of incorporating a light-weight outer dome of timber seems to have originated in the Gothic era. Wren probably observed its use in the (relatively diminutive) dome of the early seventeenth-century Parisian Church of the Sorbonne on his only-known trip

---

[17] In fact, the two original chains had burst. See Armando Schiavo, *La vita et le opere architettoniche di Michelangelo*, Rome, 1953, p. 204. The installation of additional chains, as well as the rehabilitation of the original chains, followed an investigation conducted by Giovanni Poleni in 1742. See *Memorie istoriche della gran cupola del tempio vaticano, e de' danni di essa, e de' ristoramenti loro, divise in libri cinque*, Padua, 1746.

[18] Viktor Fürst, *The Architecture of Sir Christopher Wren*, London, 1956, pp. 105-14.

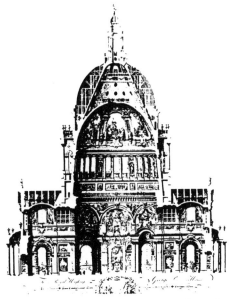

Fig.1.7. St Paul's Cathedral, London: cross-section through the great dome (J. Gwyn, 1753).

abroad, in 1665. Analogous construction can be traced back through a sequence of Venetian buildings to the thirteenth-century enlargement of the high domes of St Mark's Cathedral.[19] St Mark's light-weight lead-sheathed timber outer domes, the largest of which is about 13 m in diameter, help relieve pier footings placed on sandy Venetian subsoil that might otherwise have been overburdened if the outer domes had been constructed of conventional brick or stone—and provide still a further example of Gothic pragmatism.

CONCLUSION: GOTHIC VS. CLASSIC*
There is general consensus among architectural historians that until the pressing needs of industrial development and the introduction of new construction materials in the nineteenth century, the level of structural experimentation in European architecture never again approached its Gothic zenith. As we have already observed, this was at least partly caused by the writing down of design rules. Historians of science have extolled the flowering of technical writing in the late Middle Ages that allowed a new range of speculation: for example, Leonardo's method of 'inventing on

---

*   Apologies to Rudolf Wittkower
[19]   Mark, *Light, Wind, and Structure*, p. 147.

Fig. 1.8. Comparative cross-sections: Old St Paul's Cathedral choir and Wren's St Paul's Cathedral choir.

paper.'[20] But for the more developed technical arts such as architecture, the effect of writing seems to have been just the opposite. The inherent power of written design rules, and the publication of drawings of existing buildings accompanied by measurements, tended to codify design. Although Lechler advised his son to use his own judgement and not necessarily to follow the text in all things, the very existence of the *Instructions* must have had the effect of standardizing building form. With no scientific theory to guide experimentation, the only prudent course for assuring safe structural design would have been to follow the written text in detail.

Moreover, with the currents of fashion running northward from Italy carrying the 'new' Renaissance architectural ideals, concepts of classical style generally displaced the more pragmatic interest in devising elegant structure. This is epitomized by a comparison of the cross-sections of the thirteenth-century choir of Old St Paul's Cathedral with Wren's 1674 design (fig. 1.8). Even though the structure of the original St Paul's choir was not uncommonly tall, or light, by French High Gothic standards, it appears quite lithe compared with Wren's heavy-walled replacement. Since he was a scientist in his own right as well as Newton's colleague at the Royal Society, Wren's biographers have assumed that the extremely heavy outer walls of fig. 1.8. St Paul's must have been intended to help maintain structural stability.[21] We have established, however, that they do not.[22] On the other hand, Wren's great dome provides a rare instance from this period of design logic originating from understanding based on failure (in this case, the cracking evident in the St Peter's dome). There is no question that the style of Wren's dome derives from the Renaissance, but its structure follows more the tradition of High Gothic architecture.

---

[20] Bert Hall, '"Der Meister sol auch kennen schreiben und lesen": Writings about Technology 1400-1600 and their Cultural Implications', in D. Schmandt-Basserat, ed., *Early Technologies*, Los Angeles, 1978.

[21] This now common view seems to have been first promulgated by Somers Clark, 'Saint Paul's Cathedral: Observations on Wren's System of Buttresses for the Dome Piers and on some other things', in *Sir Christopher Wren, A.D. 1632-1723*, London, 1923, p. 73.

[22] Mark, *Light, Wind, and Structure*, pp. 155-7, 165.

Fig. 2 1. Metz Cathedral. West flank (ecclesiastical north). Five northern bays ca. 1215- ca. 1320. Three southern bays c. 1320 -90. North transept begun 1486.

2

The Rayonnant Gothic Buttresses at Metz Cathedral

Sergio Sanabria and Kristina Luce

INTRODUCTION

The cathedral of Metz is among the largest Gothic churches in Europe. (fig. 2.1) It is almost as tall as Amiens, with an average height of 41.3 m, and has an exceptionally broad nave whose average width to pier centerlines is 15.604 m.[1] Its great scale, huge clerestory windows, and unusually lightweight structure make it one of the most remarkable French buildings of the thirteenth century. (fig. 2.2) Despite this, it has attracted limited scholarly attention outside local antiquarian circles. Its neglect is probably due to three factors. One is that after the Franco-Prussian War (1870), when the Lorraine passed under German control, few French art historians studied Metz. By the end of World War I, they had established a standard chronology that ignored it. Only one German Scholar, Franz Xavier Kraus, paid much attention to Metz during this period, although the cathedral was well-tended and remodelled by the Dombaumeisters Paul Tornow and W. Schmitz, who published useful observations on its construction. The worst is that Metz has been very badly misdated, making it appear as a very retardatory structure.[2]

The broad interpretations reported here rest on observations accumulated during an ongoing survey of Metz cathedral begun in 1984. Small groups of architectural students from Miami of Ohio have participated every other year since then. Each has contributed much effort to the work, and some have advanced it in major ways. This essay focuses on a structural survey of a vault and its buttressing undertaken by Kristina Luce in 1992-3 in her Senior Honors Thesis.

[1] The widest High Gothic cathedral is Chartres, with a nave width to pier centerlines of 16.41 m.. This value is an average of 12 measurements given in John James, *The Contractors of Chartres*, 2 vols., Wyong, NSW, Australia, 1981, I, pp. 138-9. The naves of Reims and Amiens cathedrals are much narrower, 14.65 m and 14.60 m respectively, based on nineteenth-century surveys summarized in Hans Reinhardt, *La cathédrale de Reims*, Paris, 1963, p. 218, and Maurice Crampon, *La cathédrale d'Amiens*, Amiens, 1972, p. 29.

[2] The most important monograph is Marcel Aubert, ed, *La cathédrale de Metz*, 2 vols., Paris, 1931. The only earlier French study of the Gothic architecture of this region is Camille Enlart, 'Les traditions architecturales du pays Messin', *L'Austrasie* 4 (1906). Jean Bony in *French Gothic Architecture of the 12th and 13th Centuries*, Berkeley, 1983, p. 394, following too literally Pierre Marot's and Jean Vallery-Radot's articles in Aubert, dates the beginning of construction to 1257, about forty years too late.

Fig. 2.2. Metz Cathedral. Nave interior.

Fig. 2.3. Metz Cathedral. Western (ecclesiastical north) aisle from the fifth bay. The influence of Jean d'Orbais can be seen in the form of the piers and capitals of the main arcade and in the passage at the base of the aisle windows.

Under bishop Conrad de Scharfeneck (1213-24), a High Gothic cathedral arose upon the demolished nave of a tenth-century Ottonian cathedral whose unconventional north-south alignment followed the terrain slope. Construction proceeded for about ten years before war and civil disorders under bishop Jean d'Apremont (1224-38) interrupted it. Work, now in the Rayonnant style, resumed no later than 1257 and probably as early as 1250 under bishop Jacques de Lorraine (1239-60).[3]

The Metz builders came from Reims, and brought the innovations associated with the architect Jean d'Orbais, who likely participated in the project.[4] The Rémois influence is evident from the Champenois passages on the side aisles, bar tracery windows, cantonnated piers with continuous

---

[3] Little documentation exists on thirteenth-century construction at Metz; almost all that is currently known was compiled by bishop Jean-Baptiste Pelt in his monumental *Études sur la cathédrale de Metz,* 3 vols, Metz, 1930-37. The date for the beginning of work can be surmised from two pontifical Bulls of Honorius III dated 2 December 1220, allowing vacated rents to revert to the fabric of the cathedral for ten years, as well as granting indulgences to those who give alms for the construction, which was incurring heavy expenses (*vestre fabrica graves sumptus exposcat*). See Pelt, *Études,* I, 'Textes extraits principalement des registres capitulaires (1210-1790)', pp. 1-2. On 3 January 1257, Pope Alexander IV ruled that henceforth all vacated Messine canonical rents revert to the fabric of the cathedral for one year, until the very costly work be completed. This suggests a new construction initiative. I have dated stylistic ruptures based on careful observations of the building, estimating the construction time needed to complete various stages of the work, and connecting these in turn both to local events and architectural developments in the Ile-de-France and Champagne. The most complete history of Gothic Metz is Jean Schneider, *La ville de Metz au XIIIe et XIVe siècles,* Nancy, 1950. There are few modern studies of the bishopric. See Monique Arveiller-Ferry, *Catalogue des actes de Jacques de Lorraine, évêque de Metz (1239-1260),* Metz, 1957; F. Bienemann, *Conrad von Scharfeneck, Bischof von Speier und Metz und kaiserlicher Hofkanzler (1200-1224),* Strasbourg, 1887; Dom Jean François and Dom Nicolas Tabouillot, *Histoire de Metz,* 7 vols, Metz, 1769-1790, reprinted Paris, 1974; and Bishop Meurisse, *Histoire des evesques de l'église de Metz,* Metz, 1634.

[4] The master masons of Reims cathedral were named in the now-destroyed labyrinth on the nave floor. They were Jean d'Orbais (1210-28 or 1231), Jean de Loup (1228-44 or 1231-47), Gaucher de Reims (1244-52 or 1247-55), Bernard de Soissons (1252-87 or 1255-90), and Robert de Coucy (1287-1311 or 1290-1311). Jean d'Orbais directed the Reims workshop throughout the span of the first campaign in Metz. His earliest attributed work is the abbey church at Orbais, c. 1200-20, combining an Early Gothic structural system with a three-storied elevation, vertical binding of clerestory and triforium, and horizontal moldings running over the vertical shafts of the crossing piers, all features of Reims cathedral. Another attributed work is the abbey church of Essômes, begun 1205-10, where bar tracery was developed in the eastern chapels of its transept. See Robert Branner, 'Historical Aspects of the Reconstruction of Reims Cathedral', *Speculum* 36 (1961) 23-37; Branner, 'Jean d'Orbais and the Cathedral of Reims', *Art Bulletin* 43 (1961) 131-3; Pierre Héliot, 'Deux églises champenoises méconnues: les abbatiales d'Orbais et d'Essômes', *Mémoires de la Societé d'Agriculture, Commerce, Sciences et Arts du Département de la Marne* 80 (1965) 97-112; and John James, *The Traveller's Key to Medieval France,* New York, 1986, pp. 205-8, 231-4.

capitals, and classical balance of verticals and horizontals. (fig. 2.3) Much of the nave arcade up to the base of the triforium was initially completed in two distinct phases, beginning with the five southern bays (ecclesiastical west). As at Reims, the intended High Gothic elevation would have had a clerestory roughly equal in height to the nave arcade, and a triforium one third as high.

Around 1250-57 another Rémois master, possibly Gaucher de Reims, drastically altered the elevation of the nave to keep up with the contemporary soaring cathedrals of Amiens, Beauvais, and Cologne.[5] An obligatory aspect of this new Rayonnant design was to articulate shafts to bind triforium and clerestory vertically, as established around 1230 at St-Denis. The typical 3:1:3 ratio of nave arcade, triforium, and clerestory elevations of High Gothic churches was elongated to an unconventional 3:1.8:3.8. (fig. 2.4) Thus the clerestory became one-fourth taller than the nave arcade and the triforium almost twice as tall as a conventional one. Even as the height increased, the depth of the upper structural members decreased. The outer pier buttresses west of the nave are the lightest of any major French Gothic cathedral, as is the ratio of supporting stone volume to void. (fig. 2.5)

The earliest freestanding outer pier buttress is the fourth from the crossing on the west flank (ecclesiastical north).[6] Its depth at ground level is 4.498 ± .004 m, yielding a ratio of depth to nave clear span of 1:3.1. The '1250s' master dramatically reduced its upper level depth to 2.691 m, for a ratio to clear span of 1:5.19. The corresponding eastern (ecclesiastical south) outer pier buttress, by a slightly later master, perhaps Bernard de Soissons, is 3.20 m deep. Its ratio of depth to clear span is 1:4.36. The difference between western and eastern pier depths is .51 m, adjusted in the thickness of the walls of the west and east clerestories. (fig. 2.6)

The clerestory saw three major design stages, the first two corresponding roughly with the two upper buttress designs. The earliest, begun ca. 1250-57, is seen in the Tour du Châpitre and in the lower third of the western clerestory, where internal and external layers of tracery extended upwards the double wall of the triforium. The second design may be associated with the short episcopacy of Philippe de Florenge (1261-64). Philippe resigned as bishop but remained as a canon in charge of the cathedral

---

[5] Throughout the second half of the thirteenth century, innovations initiated in Reims— traceried outer pier buttresses with tabernacles, double flyers with variant angles of attack, layered tracery walls, drapery friezes, glazed tympana, and tracery motifs—are reflected immediately at Metz, suggesting the ongoing involvement of Reims master masons. The arrival of motifs from the Ile-de-France is more belated.

[6] Construction above the aisles began with the west (ecclesiastical north) tower, the Tour du Châpitre, on the fifth bay. Its buttresses are almost exact but elongated copies of those of Reims cathedral.

Fig. 2.4. Metz Cathedral. Triforium and clerestory of Bays 3-5, western flank, showing the extreme lightness of construction characteristic of the '1250s' master.

Fig. 2.5. Metz Cathedral. Outer pier buttresses above western aisle (ecclesiastical north). Note the large open spaces between the buttresses.

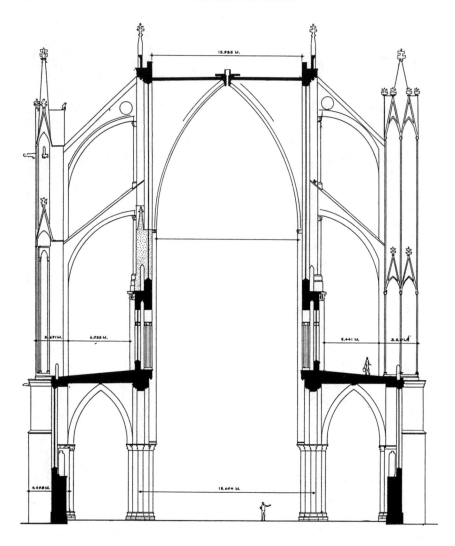

Fig. 2.6. Metz Cathedral. Sectional diagram of fourth bay (ecclesiastical north). The left half shows the design of the '1250s' master, among the slenderest in Gothic architecture; the right half, by the '1260s' master, has deeper pier buttresses.

works until his death in 1297. The 1260s' master eliminated the layered system, simplifying and lightening the clerestory wall. He began the eastern clerestory with this new design, and truncated the double wall in the west 8.14 m above the base of the clerestory (the tracery windows were truncated at 5.87 m). (fig. 2.6) The change reduced the cost of the double wall system and the deep shadows it would have cast. The 1260s' master completed only the fourth bay on the west (ecclesiastical north). Later in the century another master completed all other high windows with complex polyfoil tracery. (fig. 2.4)

The last of the five vaults of the nave must have been completed under bishop Reginald de Bar (1302-16). No extant document records a dedication ceremony. The evidence for this date is indirect, namely several illuminations in the Pontifical of Reginald de Bar (Cambridge, Fitzwilliam Museum, ms 298) representing the bishop dedicating a large five-bayed structure.[7] The extensive instructions for a major ceremony dedicating a church suggest that the consecration of the nave was imminent.[8]

The combination of vertical swelling, structural attenuation, and elaborate upper tracery is particularly striking from the exterior, where the enormous windows and thin structural frame suggest a vastly enlarged reliquary chapel, a monumental version of the Sainte Chapelle in Paris. The local name of the cathedral, *la lanterne de Dieu*, reflects this vision.

BUTTRESS STABILITY

The attenuated upper buttresses of Metz cathedral begun by the '1250s' master are among the slenderest in Gothic architecture, and have no appreciable deformations. Their technical virtuosity interested us in determining their safety factor, i.e., how closely they approached structural limits. To answer this question it was necessary to calculate the thrusts imposed on the buttresses by the vaults that they support, and this in turn demanded a full survey and analysis of a high vault. Before considering these results some general structural principles should be reviewed.

Three major component forces act at the springings of vaults: 1) the downward weight of adjacent quadrants of the rib vault; 2) a horizontal force against the adjacent bay along the axis of the wall; and 3) an outward,

---

[7] See E.S. Dewick, *The Metz Pontifical, A Manuscript Written for Reinhald von Bar, Bishop of Metz (1302-1316), and now belonging to Sir Thomas Brooke, Bart., F.S.A.*, London, 1902.

[8] Pelt interpreted an entry of 17 June 1359 in the accounts of the Metz cathedral chapter as demonstrating that high vaults were not yet completed. The text enjoins whoever may be master of the works to *faire refaire les votes*, among other tasks. This passage probably refers to the vaults of the three western bays, which were begun long after the five of the nave were completed.

horizontal force or thrust against the clerestory wall.[9] Given the strength of
stone the weight causes a comparatively small compressive stress in the
supporting masonry. After completion of construction the longitudinal force is
usually neutralized by the equal but opposite force from the adjacent vault. To
withstand thrust, however, requires either wall buttressing or flyers and outer
pier buttresses. High Gothic master masons recognized that flyers should meet
the thrust at the haunches of the vaults, about one third up along the arc above
the springings, and placed a surcharge above the *tas-de-charge*, the latter a
horizontally coursed inverted pyramid above the springing. If successfully
designed, this rigid mass transfers any potential instability to the outer pier
buttress.

The most efficient buttress would be very slender and inclined at an
angle to meet the combination of thrusts and dead loads in simple
compression along the axis of the buttress. In 1902 Julien Guadet proposed
such a buttressing system as part of a theoretical redesign of the church of St
Ouen in Rouen.[10] Antonio Gaudí used similar buttresses at the Sagrada
Familia in Barcelona and the Colonia Güell chapel at Santa Coloma de
Cervelló. Such configurations were not used by medieval builders, so our
attention must narrow to vertical buttresses. These can fail in three ways if
unable to accommodate thrust. The first is compression or crushing failure,
and as noted before it is rare. The second, shear failure, is a horizontal or
diagonal slippage of an upper block past a lower one, preventable if friction
between stones is sufficient. The third is cracking and overturning. Cracking is
the opening of vertical or diagonal fissures on the mass of the pier.
Overturning results from the opening of horizontal joints along the inner edge
of the pier, resulting in hingeing along the lower outer corner. In each of these
two cases, failure results from the thrust causing tension in the pier, which
masonry joints are unable to resist. The most efficient Gothic buttress would
be just deep enough for the thrust barely to avoid causing tension along its
edge. The Middle Third Law is the rule that describes the conditions under
which a lateral thrust causes tension in a pier.

A pier loaded only by its own weight is compressed uniformly along
its base. (fig. 2.7, A) If a small thrust shifts the resultant force slightly off
center, the distribution becomes uneven, but the base remains entirely in
compression. (fig. 2.7, B) If thrust increases to the point that the resultant
passes through the middle third of the base of the pier, the compression

---

[9] Robert Mark, *Light, Wind, and Structure: The Mystery of the Master Builders*, Cambridge,
MA, 1990, p. 116.
[10] Robert Mark, *Experiments in Gothic Structure*, Cambridge, MA, 1982, p. 43; Julien
Guadet, *Éléments et théorie de l'architecture*, 4 vols, Paris, 1899, III, pp. 340-9.

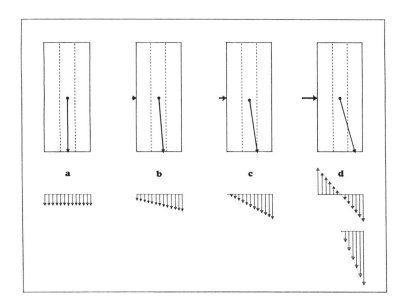

Fig. 2.7. Distribution of forces at the base of a buttress. A) Loaded by its own weight B) Loaded with a small thrust C) Loaded to the middle third D) Loaded beyond the middle third.

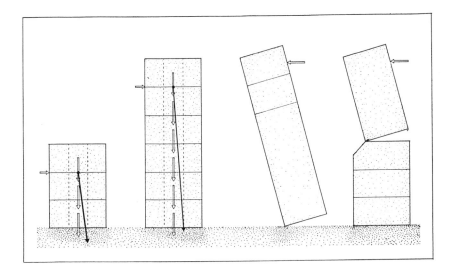

Fig.2 8: Equivalent stability of piers of different heights loaded by identical thrusts. Failure occurs at the weakest section regardless of height..

pattern becomes triangular, and the edge of the pier is no longer compressed. (fig. 2.7, C) This is the optimal limit of performance of any pier, and any additional loading eccentricity will generate tension in part of the pier. (fig. 2.7, D) This part in tension becomes ineffective due to cracking, and the remainder of the pier picks up the excess stress as additional compression. Masonry piers can withstand small amounts of this kind of failure loading, but any tensile cracking can cause further damage due to water penetration. Gothic master masons knew nothing about the Middle Third Law, but learned to examine their buildings for small crack development and to surcharge those spots to restore stable compression.

Most Gothic buttresses step out as they descend, becoming deeper and wider at the base than at the top. Contrary to intuition, such broadening does not contribute to stability. The moment arm of a pier (a measure of overturning) increases with its height, but the stabilizing weight of the pier increases proportionately. (fig. 2.8) Thus within limits imposed by slenderness, the height of a homogeneously heavy buttress does not affect the depth it needs to remain stable. If the height is raised, the increased dead load will keep the resultant at the same vertical boundary. If an unstable pier widens at any point below the lateral thrust, the result will be a higher pivoting point of failure, not greater stability. A buttress needs salient moldings or taluses only where it meets additional thrusts, but Gothic taluses seldom correspond to thrusts.

Gothic buttresses with vertical outer edges are rare. The earliest extant are at the cathedral of Bourges where the steep flyers are supported by unusual straight outer pier buttresses. It is possible that these imitate the earliest flyers of ca. 1180 at the cathedral of Paris, as reconstructed by Robert Mark and William Clark.[11] Most buttresses, such as the highly refined ones of the nave of Amiens cathedral, begun in the 1220s by Robert de Luzarches, have some structurally unnecessary taluses.

Gothic churches in the city of Metz use consistently straight vertical buttressing, beginning with the late twelfth-century pilaster buttresses of the chapel of the Knights Templars. It is possible that these were influenced by Lorrainian columnar buttresses common in Romanesque apses throughout the region.[12] At the church of St Martin in Metz, begun in the early thirteenth century, vertical buttresses with false taluses which aid in shedding water were used perhaps for the first time, establishing a local typology. The church of St.

---

[11]   William Clark and Robert Mark, 'The First Flying Buttresses: A New Reconstruction of the Nave of Notre-Dame de Paris', *The Art Bulletin* 66:1 (1984) 47-65; Clark and Mark, 'Gothic Structural Experimentation', *Scientific American* 251:5 (1984) 176-85.

[12]   See Hubert Collin, *Les églises romanes de Lorraine*, 4 vols, Nancy, 1981-86; Hans-Günther Marschall and Rainer Slotta, *Lorraine romane*, La Pierre-qui-Vire, 1984.

Vincent, begun ca. 1248 at its south absidiole, again has vertical buttresses with false taluses.[13] These are used in the mid-13th century choir of Sainte Ségolène, and the choir of the chapel of Notre-Dame-la-Ronde, ca. 1260, at Metz cathedral.

The first High Gothic master mason at Metz cathedral employed talused buttresses similar to those by Jean d'Orbais at Reims. The second master of ca. 1250-57 learned from the Messine milieu. His buttresses are not only slender but perfectly vertical on both interior and exterior edges. It is in fact possible that Metz buttressing had influenced that of Reims earlier, since the upper levels of the outer pier buttresses there, probably begun in the 1230s by Jean de Loup, exhibit a novel tentative verticality, even if compromised by their great open tabernacles backed by relatively thin masonry.

VAULTS AND BUTTRESSES AT THE CATHEDRAL OF METZ
The high vault we surveyed at the fourth bay from the crossing was either the first or second completed.[14] It springs at height of 27.82 m, is 41.13 m high at the boss, and spans 9.89 x 13.06 m.[15] The three-dimensional survey of the vault surface, and our measurement of the density of Metz stone, allowed using computerized finite element analysis to determine the distribution of forces at the springing.[16] The results of the analysis can be summarized straightforwardly. The forces on a single springing are compared below to those computed in the 1970s for a bay of Cologne cathedral by Kurt Alexander, Robert Mark, and John Abel.[17]

---

[13] Marie-Claire Burnand, *La Lorraine gothique*, Nancy, 1980; Burnand, *Lorraine gothique*, Paris, 1989; Marie-Antoinette Kuhn-Mutter, *L'église Saint-Vincent de Metz, étude historique et archéologique*, Mémoire de Maîtrise, Université de Nancy II, 1980; Kuhn-Mutter, 'L'église Saint-Vincent de Metz', *Les cahiers Lorrains* 2e trim. (1982) 131-46.

[14] The survey measured the warped curved surface of the vault. Lacking photogrammetric equipment or a high scaffold, it could only be measured from above, a task comparable to surveying a steep mountain. Debris filling the deep pits above the springing surcharges had to be hauled out before the survey could begin. Equipment included a 1-second Zeiss theodolite, steel measuring tapes, a telescoping tape, a ladder to stand on the steep slopes, and ropes to rappel down them.

[15] The spans specified are averages from the forward axes of the four main shafts. The span measured above the vaults to the face of the walls is 13.98 m. The height to the top of the boss is 42.124 m. The boss is .995 m deep.

[16] The density of the Metz stone was measured using only one sample which we had brought to the United States. It was 704 kg/m³ or 144 lbs/ft³. The finite element analysis was done by James Kipton Ping at the civil engineering firm of Steven Schaefer Associates in Cincinnati.

[17] Mark, *Experiments*, pp. 106-11.

	Thrust	Dead Load	Ratio T:DL
Metz	34,500 kg	100,000 kg	.34:1
Cologne	31,000 kg	91,000 kg	.35:1

The almost identical ratio of thrust to dead load is a result of similar geometries at both buildings. The vaults of both Metz and Cologne are very steep, which makes them very efficient.

The thrust is transmitted through the lower flyer to the outer pier buttress, and it is a relatively simple matter to compute the surcharge needed by the buttress above the point where thrust is applied in order to avoid tension. The Middle Third Law yields a value of 54,500 kg for the requisite surcharge. Given the width and depth of the buttress, the surcharge must rise 5.4 m above the lower flyer. The upper flyer springs 6.5 m above the lower, and since it and its pinnacle brace the roof against wind loads, we assume, conservatively, that under such wind-loaded conditions it would not contribute to the overall stability of the buttress.[18] If this assumption is accepted, the margin of safety for the lower flyer is 1.1 m, or 11,000 kg. Thus the western buttresses are loaded within 20 per cent of their tension limit. This is tight compared to modern engineered structures, where a 30 per cent to 35 per cent safety factor is typical.

So impressive a performance in no way implies that Messine master masons could calculate stability, but does demonstrate that through empirical observations and adjustments they could tune their building to a high order of structural refinement. Certainly we reach here a high mark in the intuitive structural experimentation conducted at the great Gothic churches of the thirteenth and fourteenth centuries, when apparently few rules were yet codified and computations were unknown. Even Jordanus Nemorarius (fl. 1220-30), who rekindled the study of theoretical mechanics, and who apparently studied the mechanics of gravity-loaded catapults or *trébuchets*, seems not to have thought of the possibility of using his own bent lever and double inclined plane theorems to analyze these vast and complex structures.[19]

---

[18] Wind loads against the original fourteenth-century roof would have been moderate, since its slope was very low, typical of medieval roofs in the Lorraine. The masons responsible for the flyers, coming from Reims, undoubtedly expected a steep roof instead of the one constructed. In May 1877 a fire destroyed the medieval roof, and the *Dombaumeister* Paul Tornow replaced it with a steep one constructed with Polonceau metal trusses. To address the higher pitch of the roof, Tornow constructed new gables on the west facade and on the transepts. Although the new roof created a substantially higher resistance to wind, Tornow did not redesign the upper flyers.

[19] Paul Chevedden, 'Artillery in Late Antiquity', *The Medieval City Under Siege*, Ivy Corfis and Michael Wolfe, eds, Woodbridge, 1995; pp. 131-76; Chevedden, Les Eigenbrod, Vernard Foley, and Werner Soedel, 'The Trebuchet', *Scientific American* 273:1 (July 1995) 66-71;

This should not be surprising. A theoretical analysis would have required both a sophisticated use of centroids, for which basic theorems were available in Archimedes and Euclid, plus an impossible knowledge of vector analysis which had not been developed yet. Leonardo da Vinci applied Jordanian principles to a correct analysis of simple three-hinged arches and their supports but his results could hardly be generalized to more complex arches. The earliest theoretical analysis of arches and their abutments, in Philippe de la Hire's *Traité de Mécanique* published in Paris in 1695, used funicular force polygons, and benefitted already from both vector analysis and Desarguesian projective geometry, yet ignored friction, again making the results essentially useless.[20] Thus the pragmatic approach of masons was the only possible route towards structural refinement available until the development of modern mechanics in the eighteenth and nineteenth centuries.

This study of the Metz buttresses may also serve to correct a misperception that the Rayonnant period, with its emphasis on linear effects, and a precious brittleness, generally disregarded structural innovation or soundness. In fact some of the most daring structures of the medieval period were designed and completed during this remarkably fertile period.

Marshall Clagett, *Archimedes in the Middle Ages*, 9 vols, Madison, 1964-80; Clagett, *The Science of Mechanics in the Middle Ages*, Madison, 1959; Edward Grant, 'Jordanus de Nemore', *Dictionary of Scientific Biography*, 7, pp. 171-79, 1973. I have not been able to find any structurally relevant parallels to Jordanus' concept of 'positional gravity', which according to Chevedden et al may reflect a study of the operation of the trebuchet.

[20] See Stephen Timoshenko, *History of Strength of Materials*, 3rd edition, New York, 1958, pp. 2-6, 62-6; Alberto Pérez Gómez, *Architecture and the Crisis of Modern Science*, Cambridge, MA, 1983, pp. 195-6, 241-3.

3

Scale and Scantling:
Technological Issues in Large-Scale Timberwork
of the High Middle Ages

Lynn T. Courtenay, FSA

INTRODUCTION: GENERAL PARAMETERS:

It is easy to admire the majestic timberwork of a great medieval hall or barn without realizing the complex decisions the carpenter had to make to create the structure we see. Yet many choices, from the selection of trees to the designing of joints, were part of this enterprise. This essay concerns some of these choices, in particular, the relationship between the overall scale of a structure and the size and shape of the individual timbers that compose its primary members. These timbers, used as tie beams, rafters, supporting posts, or struts, were once woodland trees of a certain species (traditionally oak), shape, height, and girth. Where they survive in medieval buildings, archaeological evidence suggests that trees were carefully selected with a specific structural role in mind—an hypothesis recently demonstrated for England by Cambridge botanist, Oliver Rackham.[1] While this essay concerns mainly technological issues related to structure, it also considers the broader issues of medieval carpentry traditions, woodland resources, and the wood-working process by which living trees became functional components of roofs and timber-framed buildings.

Close examination of surviving carpentry in various buildings reveals much about the economy and techniques of medieval building practice. To illustrate how the size and shaping of trees relate to structural context, three typical large-scale buildings have been chosen, each representing quite different purposes, socially, aesthetically, and architecturally. These examples

---

[1] Oliver Rackham, *Trees and Woodland of the British Landscape*, London, 1976; *Ancient Woodland: Its History, Vegetation and Uses in England*, London, 1980; 'The Growing and Transport of Timber and Underwood', in *Woodworking Techniques before A.D. 1500*, Sean McGrail, ed, *British Archaeological Reports, International Series* [hereafter *BAR*] 129, Greenwich, 1982; *The History of the Countryside*, London, 1986; *The Woods of South-East Essex*, Rochford, Essex, 1986.

date roughly to that technologically robust period between ca. 1180 and 1280. All their carpentry is oak; all have been recently dated by dendrochronology; and all are from northern Europe (England, France, and Belgium). Each type—an aisled, timber-framed barn; a high cathedral roof above stone vaults; and an open timber-roofed hall without aisles—represents a major class of building and carpentry assembly widely used throughout the Middle Ages.

Before looking at specific buildings and some of the technical demands addressed by their carpenters, it is essential to begin with some fundamental parameters. First, in all timber structures, framed or combined with masonry, span is the decisive dimension. If a space is to be spanned by a single beam or roof truss, there is a critical relationship between the usable lengths of available trees and the overall scale of the framing design. Support conditions for the timberwork are crucial engineering factors, whether the timbers must span a single width or whether the total span is divided longitudinally into aisles. If the carpenter intends to use trabeated construction with posts and cross beams, either straight or cambered, he must determine whether he can obtain straight portions of tree trunks of sufficient length and bulk. If he cannot find trees of suitable length and quality for his span, he must either alter the design of the carpentry to cope with a larger span using short members, as could be done, for example, in a multi-tiered system like the Byloke roof (see below) or in a timber arch. Otherwise he must tell his patron to reduce the width of the building—a request that might prove troublesome if dimensions have already been determined by expensive foundations and masonry—to say nothing of the patron's expectations. One hardly needs reminding of Abbot Suger's famous search under divine guidance for trees large enough to provide roof beams for the new work at St Denis.[2]

Another important concept is the ratio of span to scantling. The latter term, 'scantling', is perhaps less familiar and refers to the measured dimensions of a particular timber once it has been cut to size. Used descriptively, scantling may refer to the cross-sectional dimensions of a member in relationship to its length. Hence, carpentry of light scantling describes a structure whose members are of relatively small cross-sectional dimensions in their ratio to overall length and scale. Examples of light scantling are seen in the internal framing of the arcade posts and tie beams in the Wheat Barn at Cressing Temple (below, fig. 3.5) and especially in the tie beams of the nave roof of Notre-Dame at Paris (below, fig. 3.10). Conversely, heavy scantling refers to timbers that are thick in bulk in comparison to their length, as for example, in the heavy, molded wall posts in the great roof of

---

[2] E. Panofsky, *Abbot Suger: on the Abbey Church of St Denis and its Art Treasures*, 2nd ed, Princeton, 1979, pp. 95-6 concern Suger's arduous search for twelve great beams for the new roof.

Figure 3.1. Westminster Hall Roof, London: Detail of wall post and lower portion of the great arch, ca. 1395 (Photo, author).

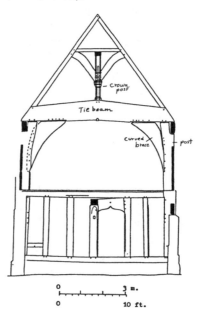

Figure 3.2. Mavesyn Ridware Manor, Staffordshire ca. 1390: Section of the gatehouse. (Courtesy of R.A. Meeson)

Westminster Hall (fig. 3.1). An even more obvious case is seen in the conspicuously enormous scantling used in the fourteenth-century gate house of Mavesyn Ridware (fig. 3.2) whose cambered tie beam, almost a meter deep at the center (36") and tapering to about 10" at the outer ends, supports a decorative crown post but spans only five and a half meters internally, thus giving a slenderness ratio of roughly 1:7.[3] Such domestic structures were clearly built to impress. Thus, by modern standards, medieval timber structures are often viewed technically as 'over built' because of the massive size of their timbers. But for what reasons? It is now an acknowledged fact that early builders in any medium did not have the mathematical means to calculate loads, bending moments, or buckling components. Rather, they worked empirically from experience—from previous successes and failures, including their own. Hence these designers, or 'engineers' could not reduce the safety factor of a structure to achieve maximum efficiency according to modern engineering principles. One assumes, therefore, that medieval builders were probably unsure and erred on the side of caution when involved with structures of very large height and span, and thus tended to be technically conservative by using more massive timbers than necessary, especially if beams were also employed to support work platforms during construction.[4] Or, a simple alternative explanation for alleged 'over-building' by medieval carpenters, suggested by Salzman, is that the prevalent use of unseasoned or green timber required 'massive timbers of most uneconomic section' to offset the inevitable warping and shrinkage.[5]

---

[3] For Mavesyn Ridware, Staffordshire see: N.W. Alcock and R.A. Meeson, *Vernacular Architecture Group Spring Conference Programme Book for 1985*, and R.A. Meeson, 'Time and Place: Medieval Carpentry Staffordshire', *Vernacular Architecture* 27 (1996) 10-24 (Fig. 4). The Gatehouse has been recently dated to ca. 1396 with a felling date of autumn 1391/spring 1392, i.e., nearly exactly contemporary with Westminster Hall whose timber was probably felled in 1393. I am grateful to Dr Alcock for bringing this material to my attention, and to R. A. Meeson for his drawing reproduced in this essay.

[4] For example, the extra hanging supports for the nave and choir roofs of Notre-Dame of Paris, discussed below, indicate that the beams were given extra reinforcement in order to sustain heavy loads. Moreover, when Viollet-le-Duc rebuilt the spire of Notre-Dame in 1858-1860, he increased the scantling of all the major structural timbers; his margin of safety, calculated from the data available, is a comfortable 8.5 times over-turning failure based on worst wind conditions. See L.T. Courtenay, 'Viollet-le-Duc et la flèche de Notre-Dame-de-Paris', *Journal d'histoire de l'architecture, nervures gothiques* 2 (1989) 53-69. This essay will be republished in English as L. T. Courtenay, 'Viollet-le-Duc and the Flèche of Notre-Dame de Paris: Gothic Carpentry of the 13th and 19th Centuries', chapter 18 in L.T. Courtenay, ed., *The Engineering of Medieval Cathedrals*, Aldershot, 1997.

[5] L.F. Salzman, *Building in England Down To 1540*, Oxford, 1952, p. 238. The prevalent use of green timber is also recorded by dendrochronologists.

But were purely technical criteria the only considerations? And was the total energy expended by man and beast all that uneconomical? Taking into account the felling, cutting to size and expensive transport, the consumption of forest resources, and the customary practice of pre-fabricated assembly, it is difficult to assess what exactly constitutes over-building in the context of social, aesthetic, constructional and structural goals. In cases where the carpentry is visible (and, therefore, usually adheres to a recognizable regional or social tradition), the status of both building and patron may well be associated with the size and physical appearance of the carpentry. In such cases, the use of large, molded timbers was likely more a matter of preference than of structural insecurity and involved a psychology of expectation and tradition. Hence, modern engineers must be circumspect in their judgment as to whether medieval timber scantling was necessarily based on overly conservative estimates of load-bearing capacity, or whether a more subtle, ascribed social value was associated with big, molded and carved oaks. These concepts of function, scale, and ornamentation are not mutually exclusive, especially in large-scale, high status buildings. For example, the huge and elaborately molded wall posts of Westminster Hall (fig. 3.1), which do take the major load of the roof down the wall, are aesthetically of equal importance to the design.[6] In short, available data from timber buildings does not support the assumption that medieval buildings were of necessity 'over built'. Indeed, when viewed in proper historical and constructional context, a number of large-scale examples of structural carpentry suggest the contrary.

WOODLAND, FOREST, AND TIMBER

An important component in any assessment of historic carpentry is the nature of woodland resources, their contemporary value, use, and management—a topic explored critically in the work of Oliver Rackham. Several points raised in his studies are pertinent to this discussion. Foremost perhaps is the intensive woodland management of both timber and underwood and the primacy given to oak (*quercus robur* and *q. petraea*) for building timber, even though oak was not necessarily the most prevalent species. Unlike the medieval forest, which may be actually unwooded, and whose nomenclature and legal status pertained to property rights and privileges, including hunting, pasture preserved for deer, and fattening pigs on acorns, woods were treated essentially as crops.

---

[6]   For Westminster Hall, see L.T. Courtenay, 'The Westminster Hall Roof: A New Archaeological Source', *British Archaeological Association Journal* 143 (1990) 95-111; L.T. Courtenay and R. Mark, 'The Westminster Hall Roof: An Historiographic and Structural Study', *Journal of the Society of Architectural Historians* (*JSAH*) 6 no. 4 (December 1987) 374-93, and by the same authors in reply to Jacques Heyman and Rowland Mainstone: 'Letters to the Editor', *JSAH* 47 (September 1988) 321-4.

Medieval management of woodland, possibly introduced by the Romans, meant that timber came from a consciously-maintained pattern consisting of coppice, underwood, and standards, at times intermingled with pasture and meadows.[7] In this system, trees that produced shoots from a stool, such as oak, beech, and hazel were routinely coppiced, i.e. cut back to the stump close to the ground to produce a cluster of trunks of moderate diameter; or, they could be pollarded, cutting the tree back to a height of five to six feet above ground; or, as in France, long trunks could be produced by regularly chopping off the lateral branches and leaving a tuft at the top. Oaks, coppiced only when in plentiful supply, were generally allowed to grow to maturity to become standards in woods, forests, wood-pasture, and hedgerows, while the surrounding undergrowth was regularly harvested as poles, viz. material for wattle work, utensils, and especially for fuel; hence the distinction between timber for major building components and wood for use on a smaller scale.[8] The importance of wood as fuel used directly or as charcoal to produce energy for ironwork, ceramics, and glass should be stressed particularly in this expansive period of the thirteenth century—a fuel hungry era when legal regulation was required to protect building timber and underwood resources.[9]

As Rackham's pioneering research on Essex woods and buildings has revealed, medieval carpenters generally used a large number of small to mid-sized oaks, i.e. with basal diameters smaller than a range of 12-15 inches (30 cm to 38 cm); conversely, builders were very sparing in using large trees, i.e. oaks larger than 15 inches (38 cm) in diameter and whose straight trunk (below branching) exceeded 25 feet (7.6 m). The extensive use of small trees

---

[7] The extensive literature on wood, timber, and woodland management includes botany, forestry, agricultural, and economic history as well as building history. There is new information about forests based on dendrochronology (see below). Of particular importance are the works of Oliver Rackham for England (above, n. 1) and also Salzman's chapter, 'Timber', in *Building in England*, pp. 210-23. See also: Cyril E. Hart, *Royal Forest, A History of Dean's Woods As Producers of Timber*, Oxford, 1966, ch. I-IV; M.G. Morris and F.H. Perring, eds., *The British Oak, Its History and Natural History*, Faringdon, Berks, 1974; and more recently Damian Goodburn, 'Woods and Woodland: Carpenters and Carpentry', ch. 6 in Gustave Milne's, *Timber Building Techniques in London c. 900-1400*, London, 1988, pp. 106-30. For continental Europe, see: Roland Bechmann, *Trees and Man, the Forest in the Middle Ages*, trans., K. Dunham, New York, 1990 and Georges Lambert, *Les veines du temps. Lectures de bois en Bourgogne*, Autun, 1992.

[8] Rackham, *Trees and Woodland*, pp. 22-3. The distribution of woodlands in England and the distinctions among woods, parks, forests, timber and underwood are particularly stressed in Rackham, 'The Growing and Transport of Timber', *Woodworking Techniques*, pp. 200-7.

[9] Hart, *Royal Forest*, pp. 43-4 records the necessity to regulate the overcutting of woods for charcoal production to support the king's ironworks in the Forest of Dean.

is evident from a number of surviving buildings, such as the Knights Templar's Wheat Barn at Cressing Temple, Essex (below).[10]

The character of the timber found in a wide variety of existing buildings indicates that medieval carpenters tended to prefer freshly cut whole trees, whereas their modern counterparts would simply go to a lumber yard and select precut boards, beams, or dowels. By contrast the medieval carpenter went into the woods to select trees of sufficient length, diameter and perhaps curvature for a specific post, beam, rafter or curved brace. Timber from trees felled after the summer growing season and before the onset of construction in early spring was, therefore, almost always worked *green* in the prefabrication stage, since oak in this form was far easier to cut and shape.[11] Since trees (preferably from local resources) were chosen by size with specific intentions, this meant that in a medieval timber-framed building or a cathedral roof practically every major member of the assembly represented an entire tree, so that the actual number of trees might well range between 400 to 700 and up depending on the design and scale of the framing.[12] The implications of this practice are manifold and reveal the close reciprocity of medieval building techniques, structural design, and the management of supplies rooted in these traditions.

CONVERSION

The process by which a woodland tree is transformed into a structural member is called conversion (fig. 3.3). The methods and tools for conversion, such as

---

[10] See below n. 22. The best recent convenient source for studies that record a variety of types of buildings is *A Bibliography of Vernacular Architecture, Volume III 1987-1989*, ed. by I.R. Pattison, D.S. Pattison, & N.W. Alcock, Oxford, 1992.

[11] On the characteristics of unseasoned oak used in the reconstruction of pagan Anglo-Saxon buildings in Suffolk and the experiential behaviour of *quercus robur* (the pendunculate oak widely used in medieval buildings), see Richard Darrah, 'Working Unseasoned Oak', *Woodworking Techniques*, pp. 219-30.

[12] Pertinent to this discussion are the published by Oliver Rackham calculating the number of trees in buildings of varied size and class listed in estimated descending building scale:

Norwich Cathedral, nave roof c. 1470 = 675 trees
King's College Chapel roof c. 1480 = 533 trees
Wheat Barn, Cressing Temple, c. 1275 = 471 ½ trees
Barley Barn, Cressing Temple c. 1205-35 = 480 trees
The Lordship Barn, Writtle, fifteenth-c. = 460 ½ trees
Monks Barn, Netteswellbury Harlow, c. 1325-1350 = 392 trees
Grundle House, Stanton c. 1500 = 332 ½ trees;
Corpus Christi College, Cambridge c. 1352-78 = 1,400 trees
Rookwood Hall Barn, Abbess Roding, AD 1539 = 206 trees

For more detail see: O. Rackham, 'Grundle House: On the Quantities of Timber in Certain East Anglian Buildings in Relation to Local Supplies', *Vernacular Architecture* 3 (1972) 3-8, esp. p. 6; and also: Rackham, 'Medieval Timber Economy', p. 89.

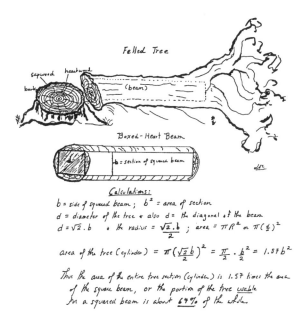

Felled Tree

sapwood   heartwood

bark

(beam)

Boxed-Heart Beam

b = section of squared beam

Calculations:

b = side of squared beam ; b^2 = area of section

d = diameter of the tree & also d = the diagonal of the beam

$d = \sqrt{2} \cdot b$ ∴ the radius = $\frac{\sqrt{2} \cdot b}{2}$; area = πR^2 or $\pi\left(\frac{d}{2}\right)^2$

area of the tree (cylinder) = $\pi\left(\frac{\sqrt{2}b}{2}\right)^2 = \frac{\pi}{2} \cdot \frac{b^2}{2} = 1.57 b^2$

Thus the area of the entire tree section (cylinder) is 1.57 times the area of the square beam, or the portion of the tree usable for a squared beam is about **64%** of the whole.

Figure 3.3. Timber Conversion: Drawing showing conversion and ratio of a square-sectioned log to a cylindrical tree.

Figure 3.4. Meaux Cathedral, S. transept, thirteenth century. Detail of bracket beneath the tie beam illustrating boxed-heart conversion; visible also are the annual rings, the medullary rays where fissures have occurred in aging, and also the heartwood/sapwood boundary. The timber at its end is only partially squared in section, and the individual irregularity of the tree can be seen, although the carpenter did shape his tree to provide three squared sides. (Photo, author)

hewing with axes, sawing, or cleaving with wedges, is technically distinguished from the type, i.e., whether a whole log, a radial section, or a plank is used.[13] In pre-modern carpentry, building timbers most often consisted of an entire tree either left partially rounded as in the bracket detail from Meaux cathedral (fig. 3.4); or frequently hewn nearly square as, for example, the vertical post in figure 3.4. This type of conversion is termed boxed heart; similarly, a tree may be halved to produce box-halved members (with timbers having a straight fair face and either partly rounded or squared sides). Hewn timbers of this type are often characterized by rippled axe marks on the surface and at times by rounded corners and waney (uneven) edges (e.g. fig. 3.4).[14]

The preference for boxed-heart or box-halved timber from whole trees and the customary use of green timber are related. It was sounder for durability and assembly purposes to use mainly the heartwood of the tree, since shrinkage from moisture loss and twisting inherent in green timber affect heartwood far less than sapwood. In fact, in a number of cases radial shrinkage for heartwood is much less than might be supposed.[15] Moreover, the hard core of the tree is less susceptible to beetle invasion and hardens with age. Yet despite sapwood's tendency toward decay, distortion, and invasion by pests, a portion of this outer wood (perhaps just a few centimeters) was often retained in construction though its presence had little effect on the actual strength of the timber as a load-bearing support.[16]

By roughly squaring timbers in an effort to use as much wood as possible, the sapwood/bark boundary or wane is partly retained and indicates that the carpenter was trying to get the maximum material from the size of tree chosen (about 64 per cent). When sapwood is present (thus preserving the heartwood/sapwood boundary) as, for example, in the thirteenth-century roofs of the Blackfriars Priory, Gloucester, where large portions of sapwood occur in the rafters, scissor braces, and collars, the carpenter has used his timber quite economically but has also incurred the known vulnerabilities inherent in

---

[13] Goodburn, 'Woods and Woodland', pp. 111-14. Rackham states that 'Tool-marks establish that timber was worked partly by sawing and partly by hewing with an adze or broad axe'. Rackham, 'The Growing and Transport of Timber', p. 208. Cf. also D.H.W. Miles and Henry Russell, 'Plumb and Level Marks', *Vernacular Architecture* 26 (1995) 33-8.

[14] *Wane* refers to the surface of the live edge of the wood beneath the bark and is important archaeologically for the imprint of tool marks and especially for establishing the felling date.

[15] On the comparative shrinkage of oak sapwood *vs* heartwood, see C.J. Venables, 'Uses of Oak, Past and Present', *The British Oak* (1974) p. 115.

[16] The differences between heartwood and sapwood are essentially chemical and concern organic materials such as oils, gums, and resins etc., that infiltrate as the wood loses water with increasing age. As its porosity decreases, the heartwood, with its blocked vessels, becomes more durable. M.G.L. Baillie, *Tree-Ring Dating and Archaeology*, Chicago, 1982, p. 54.

green sapwood, particularly the tendency to warp.[17] Thus, from observations of sapwood and the sapwood/heartwood boundary recorded by dendrochronologists, it is possible to determine how economically the carpenter used his resources. Where structural concerns and status dictated, nearly half of the wood (including most or all of the sapwood) may have been cut away to produce a neatly squared beam of uniform section. Moreover, the squaring of timber is important both aesthetically and technically, since flat surfaces and uniform-sectioned members facilitate prefabricated assembly and leveling members for assembly and jointing.[18]

In sum, the use of a large number of whole trees of boxed-heart (or box-halved) carpentry had a number of practical advantages that have a direct bearing on 'the economy' of carpentry technology. When an entire tree is used, it needs only to have its bark stripped and to be roughly hewn to size, provided the jointing is not too complex and the carpentry serves mainly a utilitarian purpose rather than an aesthetic and social one. This process saves labor and time, since it obviates the difficult process of sawing variable sizes of planks from a very large single tree. Secondly, when the tree is pre-selected according to a requisite building scale, there is little waste. Moreover, trees of small to moderate dimensions regenerate about every forty to fifty years in managed (i.e. coppiced) woods. Finally, smaller trees are easier to transport and manipulate. Close observation, then, of the actual carpentry (as opposed to reduced-scale, linear approximations of trusses) reveals a careful reckoning of scantling and function of the building components vis-à-vis resources.

DENDROCHRONOLOGY
Much of what has been learned about carpentry technology and woodland ecology has been a by-product of tree-ring dating, or dendrochronology.[19]

---

[17] O. Rackham, W.J. Blair, and J.T. Munby, 'The Thirteenth-Century Roofs and Floor of The Blackfriars Priory at Gloucester', *Medieval Archaeology* 22 (1978) 105-22. The authors report that more than half of the collars, rafters, and scissor braces have a high proportion of sapwood and wane which in some cases amounted to one third to one half of the section. To obtain this amount of sapwood from box-halved, or quartered logs, the authors suggest (pp. 113-21) these members were cut from the crowns of the large trees given to the friars by Henry III. The working of the timbers in the Priory church is however an exception, since the trees placed at their disposal were very large (c. 30' long and 2¼' in diameter). These were sawed lengthways into four to six pieces and then shaped with an adze. Ibid., p. 114. This one case thus illustrates a variety of conversion and woodworking techniques.
[18] Flat faces for joints are naturally convenient but not always obtainable when the carpenter had to cope with crooked and knotty timbers. Rackham, 'Growing and Transport of Timber', Fig. 11.4a and b, pp. 210-11.
[19] In addition to the fundamental and very readable study of Baillie (1982), see J. Fletcher, ed., *Dendrochronology in Europe: Principles, Interpretations and Applications to Archaeology and History*, Greenwich, 1978; E. Hollstein, *Mitteleuropäische Eichenchronologie*, Mainz am Rhein, 1979; G.-N. Lambert, 'La dendrochronologie à travers

Simply stated, this dating strategy arises from the ancient observation that each year trees like oak, which alternate between a growing and dormant phase, produce a seasonal ring whose width is principally determined by climatic conditions. Dendrochronology is based on matching the measured, annual ring widths from specimens of unknown date and overlapping them with a securely dated ring sequence called a master chronology. The starting point for constructing a chronological database begins by felling trees and counting their annual rings. This 'living-tree sequence' is gradually extended backwards in time as more data becomes available from samples taken from a building with a securely documented date or by cross-dating and bridging undated (floating) ring sequences—a method first developed by the American, A.E. Douglass in Arizona in the early twentieth century by physically matching specimens of yellow pine.[20] Now, using bridged ring sequences, master chronologies for oak have been constructed covering thousands of years.

Presuming that variations in moisture and temperature affect all trees of the same species similarly within a given region and altitude (i.e., a micro-climate site) average widths are quantified to produce computerized statistical databases used by a number of laboratories for cross-dating. Thus, dendrochronology is a sophisticated statistical science of dating buildings and artefacts in archaeology, climatology, and art history.

The importance of denchronology for the history of building technology needs little comment, since being able to arrange observable phenomena in chronological sequence is crucial to understanding how certain practices evolved or were retained. But perhaps less obvious to architectural historians is the important corollary regarding the economy of woodland resources and traditional building practices. For example, we can often discover the nature and quality of the trees used by carpenters, and at times, where they obtained their timber, and from what kinds of forests. Moreover, the amount of heartwood versus sapwood can also yield insights into the broader technological context.[21]

---

les laboratoires européens', *Les dossiers de l'archéologie* 39 (1979) 56-65; R.R. Laxton and C.D. Litton, *An East Midlands Master Tree-ring Chronology and its Use for Dating Vernacular Buildings*, Nottingham, 1988; and, for a wide survey of the field see the extensive bibliography in: Patrick Hoffsummer, *Les charpentes de toitures en Wallonie, Typologie et dendrochronologie (XIe - XIXe siècle)*, in *Études et Documents, Monuments et Sites*, Namur, 1995. I am especially grateful for the author's having provided me with a copy of his manuscript prior to its publication. Most recent material in this exceedingly active field appears in various symposia and journals like *Vernacular Architecture, Dendrochronologia*, and *BAR, international series*.

[20] On the early work of Douglass, see Baillie, *Tree-Ring Dating*, pp. 28-37.

[21] There are no precise rules for determining the proportion of hardwood to sapwood in European oak, since ring width and regional growth patterns make this a highly variable

AISLED CONSTRUCTION: CRESSING TEMPLE, ESSEX

Dating to ca. 1257-90, the Knights Templar's Wheat Barn at Cressing Temple Manor, Essex (figs. 3.5-3.8), is a well-preserved and technically-advanced example of timber-framed aisled construction, a common type of structure of long duration across northern Europe. As a totally 'framed' structure, the carpentry is self-sustaining and is entirely composed of jointed timbers using a bay system of construction.[22] From a technological perspective the fully-integrated framing is an important structural feature (see below). The timber frame rests on low masonry sills (a notable advance from earlier earth-fast structures); the original stud-and-plank walls have been replaced by brick, but apart from the ground sills, the entire thirteenth-century structure was conceived and built in wood.

This barn, whose architectural and historical context is well known, provides an elegant exemplar of thirteenth-century structural carpentry. Two

---

feature. Baillie, *Tree-Ring Dating*, pp. 54-60. However, we do know certain facts based on tree botany and accumulated regional data. First, the sapwood is a relatively small portion of oak, since it is known that 75 percent of the fluid transference takes place in the outermost ring so that the wood becomes less porous with age as the inner portion dries and hardens. Cf. K.A. Longman and M.P. Coutts, 'Physiology Of The Oak Tree', *The British Oak* (1974) 209. From a modern perspective, the waste factor in removing the '2-3 inches in width of sapwood' from oak is estimated at 10 to 20 per cent loss from cutting oak to sap-free specifications (i.e., considerably less than conversion to a square-sectioned beam (above, fig. 3.3), Venables, 'Uses of Oak', ibid., p. 120. Cf. J. Hillam, R.A. Morgan and I. Tyers, 'Sapwood estimates and the dating of short ring sequences', *Application of Tree-Ring Studies*, Greenwich, 1987, 165-85. The transition between the sapwood and heartwood of oak is fairly constant in a given region and can thus be used to estimate the felling date. For England a standard ring number for sapwood varies between 10 and 50 rings in London and 15 to 40 in Warwickshire; in general an average of c. 30 rings (30 years) has been used when less refined data is absent. For northern France and the Loire valley, a range of 18 to 50 rings is reported, whereas for Brittany, 9 to 28 rings of sapwood are found. Michael Jones et al., 'The Seigneurial Domestic Buildings of Brittany: A Provisional Assessment', *The Antiquaries Journal* 69 (1989, part I) 100. Recently, the calculation of sapwood has been related to actual measurement of width in millimeters as in Fletcher's earlier work, which estimates about an inch of sapwood on average and assumes a consistent ring width of 1 mm, so that a measured inch = 25 years. Even so, microclimatic conditions are highly variable and one must keep in mind the assumptions involved in using statistical averages for reality! The sapwood-width method has been recently applied to eastern Belgium by Hoffsummer. Using a database of P. Gassman in Neuchâtel, he calculates the average amount of sapwood from the last hardwood ring to be 2 cm wide, which in turn averages to 16 rings, or equal to 16 years with an additional variable +/- 5 years for the felling date. Hoffsummer, *Les charpentes*, p. 38.

[22] *Cressing Temple, A Templar and Hospitaller Manor in Essex*, London, 1993 is the fundamental study for this site. The pertinent contributions for this essay are Oliver Rackham, 'Medieval Timber Economy as Illustrated by the Cressing Temple Barns', pp. 85-92; Dave Stenning, 'The Cressing Barns and the Early Development of Barns in South-East England', pp. 51-75; and Ian Tyers, 'Tree-ring dating at Cressing Temple and the Essex Curve', pp. 77-.

aspects of this rather exceptional structure will be considered: 1) the technological advances seen in its carpentry compared to related structures (many of which are observed by Dave Stenning); and 2) the relationship between the sectional dimensions of the major timber members and the nature of the trees used in its construction vis-à-vis the building's overall scale.

The seven-bay barn, probably constructed from east to west, measures 39.75 m x 12.2 m internally; the largest span, that of the central aisle, is 6.6 m; the bay width is 5.6 m. The arcade posts (roughly 36 cm in section and 6.4 m long) are nearly as tall as the more slender tie beams (c. 25 cm in section and 6.9 m long with an aspect ratio of 1:27).The posts and tie beams are framed by normal assembly, i.e. the longitudinal arcade plate occurs between the post and the tie beam, so that the tie beam can be joined to both the post and the plate by mortise and tenon joints (fig. 3.7).[23]

The carpentry of the Wheat Barn exhibits a number of technical attributes that augment the general cohesiveness of the structure (see fig. 3.6). Specifically these are: 1) the firm link between the internal framing and the peripheral walls, achieved to a large extent by the jointing of the large-sectioned (c. 36-37 cm square) sole-plates to the aisle posts, and 2) by the presence of aisle ties (c. 25 cm square in section) braced by short, sturdy diagonal struts. 3) The framing is integrated lengthways and in the roof slope by square-set purlins (fig. 3.8) which are clasped ingeniously between the upper collar, rafter, and a short strut, thus resisting a tendency for racking (lengthways deformation). 4) The long diagonal timbers, passing braces, are aligned in two separate lengths so that shorter timbers could be used (a design factor important to the economy of construction); the lower sections of the passing braces link the outer frame post to the inner (free standing) arcade posts and are halved across the aisle ties respectively; the upper sections extend from the posts across the tie beam and collar to the opposite slope of the steeply pitched roof (above, figs. 3.6 and 3.8). This configuration forms parallel, secondary rafters and a scissor brace in the upper section of the frame thus providing considerable integration in the transverse plane (fig. 3.5). There are also sturdy diagonal braces from the arcade posts to the tie beams as well as slightly bowed diagonal braces from the aisle ties to the sole plates. Finally,

---

83. The estimated felling date range for the Wheat Barn, which contains little sapwood is 1257-90 for a felling date, see Tyers, ibid.

[23] The alternative method is called 'reversed assembly' where the plate rests on top of the tie beam. See J.T. Smith, 'The Early Development of Timber Buildings: The Passing Brace and Reversed Assembly', *Archaeological Journal* 131 (1974) 238-63.

Figure 3.5. Cressing Temple, Essex: The Wheat Barn ca. 1260: Interior from the central aisle. (Photo, courtesy F.R. Horlbeck)

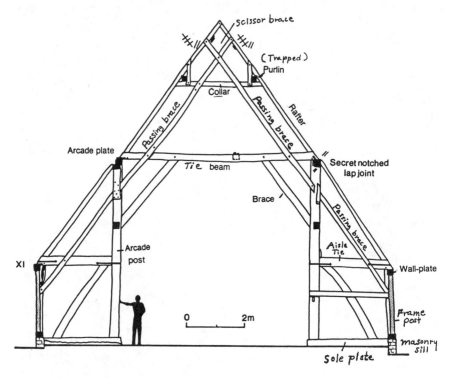

Figure 3.6. Cressing Temple Wheat Barn: Transverse section. (Drawing courtesy of D. Stenning)

Figure 3.7. Cressing Temple Wheat Barn: Detail of arcade post and plate assembly. (Photo, author)

Figure 3.8. Cressing Temple Wheat Barn: Roof detail: trapped purlin and scissor brace. (Photo, author).

it has been observed that the original arcade post heads are jowled (enlarged) at their upper end (fig. 3.7), and that 'such jowled posts generally represent upturned trees with their jowls shaped from the root bowl of the tree... [a technique which is] effective in achieving a three-way joint'.[24] Generally speaking, the jointing is sophisticated and shows a development over older forms. Two cases in point are the splayed and tabled scarf joint with a key as well as pegs, used to join timbers end-to-end, and the ubiquitous use of the secret notched lap joint, so called because the notch which functions to prevent withdrawal is hidden inside the splay of the joint.

While the carpentry per se is certainly elegant as barns go, it becomes all the more fascinating when the construction is viewed from the perspective of medieval ecology, as demonstrated by Oliver Rackham. Combining carpentry features, boxed-heart conversion, Rackham's classification of tree dimensions in Essex relative to woodland occupancy, and the quantification of the components of the Wheat Barn, we can distinguish some rather striking, but by no means atypical observations: all of the timber is oak, mainly boxed-heart timber that has been carefully squared to leave very little sapwood and few waney edges, some of which occur in the rafters. All lengthways sawing has been avoided, and the timbers have been shaped with axes. Rackham has calculated that 472 trees went into building this barn—a sum equivalent to about one third of a cathedral roof.[25] But, most critical ecologically, 54 per cent of the timber (i.e., the rafters, ca. 13 cm x 14.6 cm in section) derives from quite small trees of about 18 cm to 20 cm (7" to 8") in basal diameter. This Class I-size oak, harvested from carefully managed (coppiced) woods, occupies 1/100 of an acre for only about fifty years.[26]

The lightness of the scantling and corresponding economy in construction can be gauged also by the fact that on average at least one third to one half of the timber has been cut away during conversion, thus leaving squared members of relatively light, uniform section with little or no sapwood.[27] Moreover, the cutting away of sapwood not only produces more durable components less susceptible to pests and decay, but also the sapwood

---

[24] Stenning, 'The Cressing Barns', p. 70. Inspection of the posts appears to support this inference, though in some cases the posts do not appear jowled and have only a slight taper. Clearly, the variety in individual trees of this size had to be adapted to the carpenter's purposes.

[25] Rackham, 'Medieval Timber Economy', p. 88.

[26] Ibid., p. 86.

[27] This amount of sapwood removal is probably not true for the rafters which have an average scantling of 5" x 5.75" and thus a diagonal of 7.61". Calculations indicate that about 90 per cent of the diameter of a tree 8.4" was used, though Rackham says the rafters were converted from trees just 8" in girth. Rackham, ibid., p. 88. Cf. my fig. 3. The linear equivalent for conversion with sapwood retained is 0.71 of the scantling of the member under nearly ideal conditions of a straight and nearly untapered tree.

removal appreciably reduces the total weight of timber by about a third—a highly significant factor for the transport of balks from the felling to the erection site, especially for trees of large dimensions not always locally available. The combined removal of sapwood and pre-selection of trees of requisite size are thus significant practical factors in the technology and economy of medieval timber construction.

The second largest timbers in the Wheat Barn, Rackham's Class II, probably from local sources, come from trees, ranging in diameter from ca. 23 cm (9") to ca. 32 cm (12.5"). By calculating the average tree diameter and estimating a conservative factor of 0.65 for conversion, the average scantling for timbers in this range (e.g., the diagonal braces and longitudinal purlins) is about 17 cm to 18 cm (ca. 7") and hence just slightly larger in section than the rafters. These members represent oaks up to one hundred years of woodland occupancy, i.e., twice that of the smallest trees. Significantly, the timber from both Classes I & II comprise 88 per cent of the total trees employed in the barn.[28]

Timbers of more exceptional dimensions (the largest diameter calculated by Rackham was ca. 59 cm) comprise only 12 per cent of the timber and are judiciously allocated to major structural members such as the arcade posts and the sole-plates discussed above. From this information, it can then be estimated that the largest timbers after conversion are roughly 36 cm square. But these large trees are only 24 in number and, although centuries old, they represent quantitatively only a small portion of actual forest resources. The longest single timber discovered (an arcade plate) is 46' long and spans two and a half bays, thus providing axial integration at the mid point of the structure. This length is exceptional, however, since the long diagonal passing braces are not single members but rather two timbers placed end to end so that shorter trees could be used.[29] One could multiply similar observations and calculations, but I think this example makes clear the economical use of timber scantling corresponding to structural function. The carpentry of the Cressing Wheat Barn thus illustrates the interaction of design, materials, constructional practice, and economy—a range of technological concerns that include woodland management, transport, overall costs, and the ultimate success of a structure of elegant and efficient proportions. In light of these interactive factors, we can begin to appreciate similar concerns and other technical challenges in much larger-scale buildings.

---

[28] My percentages and scantling calculations are derived from Rackham's figures.
[29] Rackham, ibid., p. 89.

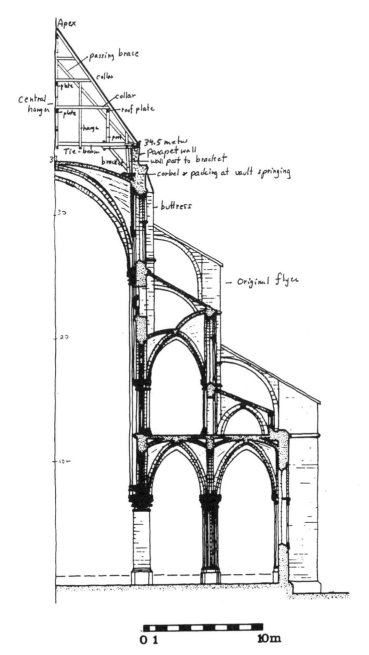

Apex

passing brace

collar

plate

collar

Central hanger

roof plate

plate

hanger

post

34.5 metw

Tie-beam

parapet wall

bracket

wall post to bracket

corbel & packing at vault springing

buttress

Original flyer

O 1 10m

Figure 3.9. Notre-Dame Cathedral, Paris: Reconstruction of original nave prior to 1240 by W. Clark and R. Mark, with roof framing after Deneux and Ostendorf with some additions by Courtenay

NOTRE-DAME-DE-PARIS, THE NAVE ROOF

In turning from barns to cathedrals, specifically to Notre-Dame in Paris, considerably different technological challenges arise for builders that nonetheless reflect similar choices on the part of the master carpenter as to design, woodland resources, and the achievement of stability. Here, in dramatic contrast to the ground-level, timber-framed Essex barn, the task involved constructing a vast cathedral roof above high masonry vaults. At Notre-Dame (fig. 3.9), the timberwork is seated on the daringly thin parapet wall (only 76 cm wide and 35 m above the ground) and surmounts the tallest Gothic structure of the late twelfth century.[30] Because of the unprecedented height of Notre-Dame, both masons and carpenters had to cope with forces not just of gravity but also wind on this tall, thin structure.[31]

Consequently, the master carpenter of the nave of ca. 1220 may well have been the first designer to achieve effective longitudinal bracing in a large-scale roof during this formative phase of Gothic technological development. He was clearly concerned with the roof's stability relative to the masonry support conditions of the upper wall and with anchoring the tie beams in order to provide sturdy work platforms for vault construction, since the roof was erected before the vaults. Building at such a height, the carpenter of the Paris nave, no doubt familiar with similar conditions in the newly-completed choir of 1182, was acutely aware of the need for length-ways rigidity and wind-bracing at various levels for a roof. These two critical factors, namely, the support conditions at the base of the roof, i.e., at the tie beams at the wall head, and the development of longitudinal stability, received particular attention in the design.[32]

---

[30]   The precise height of the nave parapet has been scaled from a number of sections of Notre Dame, most recently that of Clark and Mark reproduced in figure 8. My measurement of the wall thickness was taken at the fifth bay of the nave through the wall (58.4 cm) with the addition of the 'jowled' masonry overhang of 17.7 cm.

[31]   See W. Clark and R. Mark, 'The First Flying Buttresses: A New Reconstruction of the Nave of Notre Dame de Paris', *Art Bulletin* 66 (March 1984) 540-69. On the early Gothic Cathedral of Paris, cf. the contribution of Robert Mark in this volume.

[32]   Longitudinal stability and axial integration are of major significance in the history of roofing technology, especially for the expansive building endeavours of the thirteenth century. See Henri Deneux, 'Évolution des Charpentes du XIe au XVIIIe siècle', *Architecte* 4 (1927) 49-68 and the more recent summary in L.T. Courtenay, 'Timber Roofs and Spires', in *Architectural Technology up to the Scientific Revolution*, ed. R. Mark, Cambridge, 1993, pp. 214-16. The dating of the choir of Paris is documented by two texts, the account of Robert of Torigny of 1177, who states that the choir was complete except for the roof (*tectio*) which Mortet translates as *comble,* and the record of the consecration of the High Altar 19 May 1182, by the papal legate, Henri de Château-Marçay, and Bishop Maurice de Sully. Mortet, pp. 43-5. This interval of time allowed four summer seasons for completing the upper framing, battens, and leading of the choir roof. With the construction of the nave already in

Figure 3.10. Notre-Dame, Paris, Nave Roof: Model by Henri Deneux in the Centre de Recherches sur les Monuments Historiques, Paris. (Photo, author)

Figure 3.11. Notre-Dame, Paris, Nave Roof: View of interior above the tie beams illustrating the rafter couples, major trusses, and the posts that carry the longitudinal side plate. (Photo, author)

The timber scantlings of the nave whose carpentry dates typologically between 1180 and 1250,[33] displays a similar use of boxed-heart or halved oak (not chestnut as rumored) as well as a highly sophisticated structural integration of closely-spaced rafter couples (ca. 14 cm square) of light, uniform scantling spaced 64 cm apart from axis to axis.[34] The single, main-span trussed roof contrasts with the multiple spans of aisled, bay construction of Cressing Temple barn. It is significant that the carpenter of Notre-Dame limited his tie-beam trusses to every fifth rafter couple (figs. 3.10 & 3.11). As a result, these major trusses are spaced at intervals of just over three meters, thus dividing the roof into longitudinal bays marked by the more complex tie-beam trusses that nonetheless still have the same light scantling as all the other rafter couples (fig. 11). Moreover, by reducing to eleven the number of long tie beams, the largest-sectioned members in the roof, the carpenter has decreased considerably the consumption (and expense) of very large oaks. Excluding the ties, the economy is particularly evident from the measured dimensions of other roof members whose individual maximum depth of a nearly square section does not exceed 18 cm (7") after conversion. Indeed, the scantlings are quite uniform on average and range from about 13 to 18 cm. If one uses the same (conservative) conversion factor as previously (0.65), the average size girth of the trees for the roof of Notre-Dame is about 27 cm (10.9"), i.e., a size that corresponds to Rackham's Class II tree, occupying 1/100 acre for about one hundred years.[35]

In relationship to the scale of the roof, with its average base tie of 15 m long, a height of 9.5 m and pitch of ca. 56 degrees, the timber scantlings are exceedingly light. Although the square-hewn tie beams extending to the outer parapet wall and spanning the width of the nave (varying east to west from 13 to 14.5 m internally) may appear massive to modern eyes, the sectional dimensions of these members is only 25 cm deep x ca. 24 cm wide after conversion. Hence, these rather exceptional beams (possibly achieved by the practice of lopping off lateral branches) derive from very tall oaks about 38 cm in girth. These trees fall into Rackham's Class III size representing four units of woodland occupancy for about 200 years.[36] From another perspective, however, if one compares the depth of the tie to its overall length, the slenderness ratio at Notre-Dame is 1:60 in comparison to the analogous ratio (1:27) of the light proportions of the Cressing Temple tie beams. Also, it

---

[33] The dating of the *charpente* of Notre Dame is in progress by Virginie Chevrier under the direction of G-N. Lambert and P. Hoffsummer, Laboratoire de chrono-écologie, Besançon, where preliminary results have yielded dates for the nave tie beams from the twelfth century extending to 1275. (V. Chevrier, personal communication, 1995).

[34] Deneux, 'L'Évolution des charpentes', p. 58.

[35] Above, Rackham, 'Medieval Timber Economy', p. 86.

[36] Rackham, ibid

should be remembered that these beams carried work platforms and presumably heavy equipment and materials. Similar to the Cressing Wheat Barn and to ecclesiastical roofs of the thirteenth century in general, the proportion of the biggest members to the most numerous components roughly correspond. Even so, Notre-Dame's structure is both larger in overall scale and lighter in scantling, resulting in an elegant structure of technological finesse above the highest six-part vaults of the twelfth century.

The substantial support engineered at the base of the nave roof of Notre-Dame clearly indicates the master carpenter's concern for seating his lightly-framed roof. To provide the desired support conditions, the top course of masonry at the wall head was enlarged on its inner face in the same way a supporting post might be jowled in timber construction. To achieve such an integrated structure, carpenter, mason, and smith no doubt worked closely together to insure the seating for the roof, parapet wall, and the lead-covered guttering valley on the exterior.[37] In addition to trenching his tie beam over two wall plates (spaced 24 cm apart), the carpenter added a third, inner plate (ca. 17 cm x 14 cm) that oversails the wall head on the interior and is firmly braced from below by a bracket structure with three diagonal struts and wall post supported on a masonry corbel (fig. 3.12a).[38]

Evidence that the tie beams were designed to sustain extra loads during construction is manifested in the carpenter's incorporation of no less than five supports along the total length of the internal tie-beam span (ca. 12.5 to 13 m): one at each end with a strong bracket (see above) and three others at points across the central span. Originally, three long hanging clasps in tension (misleadingly called 'king posts' and 'queen posts') supported the ties from above and below, thereby preventing any potential sagging caused by work platforms and machinery (fig. 3.12b). Today only the central hangers and those on the north side remain *in situ*, and as the reinforced ties are presumed to function as a tension tie in a truss, engineers would likely find this to be an example of 'over building' since the stresses in the tie are probably far less than present load conditions warrant. We will never know how these beams were originally loaded, but we do know that medieval builders had no means to calculate bending stresses absolutely. Yet the hangers themselves reveal considerable perception and elegance of a design for a member in tension.

These double-hangers, or clasps, are formed by two very slender polygonal timbers, chamfered along their length; they are larger and

---

[37] In 1196, Maurice de Sully, bishop of Paris, left 100 pounds to the cathedral of Paris for the leading of the roof. Mortet, 'Étude Historique', pp. 45-6.

[38] L.T. Courtenay, 'Where Roof Meets Wall', *Science And Technology in Medieval Society*, ed. Pamela Long, New York, 1985, 109-17.

Figure 3.12. Notre-Dame, Paris: Nave Roof: a) main truss with tie beam and lower bracket; b) detail of upper portion of the lateral tension clasp (hanger) and the dowel across the second transverse collar. (Photos, author)

rectangular in section (ca. 17 cm x 10 cm) where they enclose the tie beams and two collars respectively. The two parts of the hangers are joined by a single dowel above each collar whereby the dowels form a short cross-bar from which the hangers are suspended (figs. 3.10 & 3.12b). In addition the vertical hangers are cut to fit over the horizontal collars. They then extend below the tie beam and are secured by larger wooden dowels underneath the ties whose weight will naturally keep the hanging clasp in tension. While this technique of parallel hangers to reinforce the tie beam appeared earlier in the choir roof of Paris and in the thirteenth-century roofs of Notre-Dame at Mantes and the western portion of the nave of Rouen cathedral, none of these exhibits the slender proportions, chamfered timbers, and technical sophistication of the master of the Paris nave.

Another example of the economy and ingenuity of the Paris carpentry is the dual function of the central tension clasp, which, stabilized by diagonal passing braces, both sustained the mid-point of the tie-beam and carried length-ways roof plates at two levels (fig. 3.10).[39] These plates (somewhat like purlins in later roofs) link the 'bay' trusses axially at the mid point of the roof. Thus, the concern for longitudinal stability against racking (seen also in the Wheat Barn) contributes significantly to the success of the Paris nave framing where high winds likely produced difficulties in the achievement of stability in such a tall structure.

Although there are technically no side purlins in the plane of the rafters, lengthways side bracing is provided in the lower half of the framing by the incorporation of square-set longitudinal plates carried on sturdy posts (18 cm square and in compression rather than tension, fig. 3.11). These posts are analogous to earlier aisle posts and are tenoned into the tie beam above the strongly-braced saddle bracket underneath the tie (fig. 3.12a). This assembly, moreover, functions in several ways to sustain any applied load from the post and roof plate and also to anchor this light-sectioned framing firmly to the

---

[39] The exact number of plates per truss carried by the central hanger is unclear from the historical evidence recorded at different stages of preservation. For example, the section of the crossing reproduced by Emile Leconte (1841) shows four plates, one at each of the three transverse collars and at the ridge, but no lateral hangers. See E. Leconte, *Choix des monuments du moyen âge. Notre-Dame de Paris*, Paris, 1841, pl. 60. The earliest measured drawing of the Paris nave roof, to my knowledge, is found in Jean-Charles Krafft, *Traité sur l'art de charpente théorique et pratique*, IV, Paris, 1821, p. 4 , pl. X. Krafft's section clearly shows no ridge piece and only two plates framed into the central hanger. This is similar to that of Ostendorf (1908) and Deneux (1927), although Ostendorf's section shows a longitudinal member just beneath the apex, F. Ostendorf, *Geschichte des Dachwerks*, Berlin and Leipzig, 1908, p. 20. Cf. Deneux, 'L'Évolution des charpentes', Fig. 87, p. 58 and Deneux's Dessin No. 6989 (1915) in the archives of the Centre de Recherches sur les Monuments Historiques, Paris.

wall head—again an example of the technological inventiveness of the master carpenter of Notre-Dame.

BYLOKE HOSPITAL WARD

From the realm of Quasimodo we descend to the final example of thirteenth-century structural carpentry, the extraordinary roof of a ground-level open hall of the Hospital at Byloke, Ghent (fig. 3.13). The oak roof of the main hospital ward (*Ziekenzaal*) is unquestionably one of the most important survivals of monumental carpentry of the Middle Ages and has recently been dated by dendrochronology to 1251-55.[40] This infirmary hall is also a striking example of a particular class of medieval roofs mostly found in England and Flanders that span a space without interior supports or aisles.

Clearly there are major conceptual and structural differences between the Byloke roof and that of Notre-Dame of Paris some fifty years earlier. Firstly, the Byloke roof is an open roof meant to be seen, not concealed above masonry vaults. Structurally, there is a difference in the use of double framing, formed by heavy-sectioned structural timbers called principal rafters (or principals) at long bay intervals (fig. 3.14). These principals alternate with dramatically smaller-sectioned, common-rafter couples (seven per bay with the central frame emphasized at its projecting 'ashlar foot'). Scantling here indicates load-bearing members and bay divisions (fig. 3.15), though these impressive molded principals are equally important for design and status, perhaps reflecting visually the patronage of Jeanne, Countess of Flanders.[41]

In this kind of structural system, ultimately derived from the aisled and bay construction seen earlier at Cressing Temple, the loads of the roof are transferred to the principal frames by longitudinal plates and purlins set in the plane of the roof slope and carrying common rafters (fig. 3.14). Because the aisles have been eliminated, the Byloke roof belongs typologically to a large family of structures called aisled derivatives, where, for social and aesthetic

---

[40] See in particular, Patrick Hoffsummer, 'La Charpente de la Salle des malades de l'Hôpital de la Biloque à Gand', *Actes du 51e Congrès de la Fédération des cercles d'archéologie et d'histoire de Belgique*, I, Liège, 1992, 94-5; M.C. Laleman et P. Raveschot, 'L'Hôpital de la Bijloke à Gent: Premier bilan de la recherche archéologique', in ibid., II, Liège, 1994, pp. 129-35. Guido Everaert, M.C. Laleman, and Daniel Lievopis, 'De Tweede Ziekenzaal Van De Bijloke', *Stadsarcheologie bodem en monument in Gent* No. 4 (1993) 5-23. Cf. also Hoffsummer, *Les charpentes de toitures.*

[41] The hospital of Byloke was transferred to its present site in 1228. Donations to the convent began ca. 1233 and in 1243 the countess among others undertook to increase the size of the community. L. van Puyvelde, *Un hôpital du moyen âge et une abbaye y annexée: La Biloke De Gand*, Ghent and Paris, 1925, 19-21. The primary work for the documentary history of Byloke remains J. Walters, *Geschiedenis der zusters der Bijloke te Gent*, 2 vols., Ghent, 1929-30. A recent study of the sources and the archaeology is, Lalleman and Raveschot, 'hôpital de la Bijloke', pp. 131-35. Cf. above n. 41.

Figure 3.13. Byloke Hospital, Ghent: East view of the interior prior to its restoration. (Photo courtesy of Dienst Stadsarcheologie, Ghent, M.C. Laleman, Urban Archeologist)

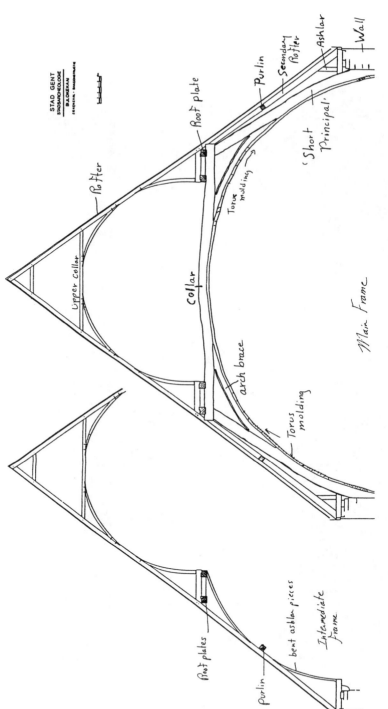

Figure 3.14. Byloke Hospital, Ghent, Roof: Transverse sections of principal rafter frame and the intermediate common rafter. (Drawing courtesy of M.C. Laleman, Urban Archaeologist, Ghent)

Figure 3.15. Byloke Hospital, Ghent, Roof: Upper section revealing the trefoil design. Note the difference between the principal frames and the intermediate light frames as well as the longitudinal plates and side purlins. (Photo, author)

reasons, internal arcades and posts have been removed in favor of one vast interior space. A variety of buildings bears witness to carpenters' efforts to contrive ingenious and attractive means of achieving a large span without internal arcades. For reasons of stability and support conditions, the heavy principals typical of these large thrusting roofs like Byloke, extend below the wall head and are carried on masonry corbels, which at Byloke were re-modeled in the seventeenth century (fig. 3.16). The evolution from aisled to unaisled interiors is most clearly preserved in the experimental roof forms discovered at Pilgrims' Hall, Winchester (dating to 1308).[42] Indeed, the creation of the large-span hall without aisles is one of the great technological triumphs of the second half of the thirteenth century.

The roof at Byloke, remarkable for its scale and nearly a half century earlier than Pilgrims' Hall, demonstrates the carpenter's masterful use of curved heavy principals (termed a roof with short principals) to form a lower frame which supports a superstructure. This multi-tiered design resembles closely the central upper hall of the Hospital of St John in Bruges traditionally dated to ca. 1290. Both at St John's and at Byloke, the gable walls characteristically reveal the vestige of aisled construction in the placement of the wall brackets that support the ends of the longitudinal plates.[43]

The recent dendro-dating of Byloke to the thirteenth century will no doubt have important ramifications for future interpretations of the technological developments in monumental roof carpentry, since it has been generally assumed that works of such prodigious scale surely belonged to the mid-fourteenth century or even later. Among unaisled medieval hospital wards, Byloke's large internal span of about 16 meters (15.92 m or ca. 53') is exceeded by only several meters in the timber-vaulted and panelled roof of the hospital at Tonnerre in Burgundy dated to 1295.[44]

---

[42] J. Crook, 'The Pilgrims' Hall, Winchester, Hammerbeams, Base Crucks and Aisle-Derivative Roof Structures', *Archaeologia* 109 (1991) 129-59.

[43] In this context see N.W. Alcock and M. Barley, 'Medieval Roofs with Base Crucks and Short Principals', *Antiquaries Journal* 52 (1972) 144.

[44] The width given in different accounts ranges from 18.6 m internal span to 21 m and the length 88 m (61' x 288')—a size roughly comparable to the largest medieval hall at Westminster Palace, which was converted to its un-aisled form in 1395 and measures 67.5' x 239.5' internally. The infirmary at Tonnerre predates by a century Westminster Hall and remains the largest timber vault in northern Europe. On the *charpente* of Tonnerre see Deneux, 'L'Évolution des charpentes', p. 60 who found the tie beams to measure 21.25 m in total length and 0.3 m square in section. The rafters are single timbers: 19.25 m long and 0.16 m x 0.18 m in section at the ridge and 0.20 m x 0.26 m at the wall plates. As at Notre-Dame, the ties with hanging posts occur at every fifth frame with four rafter couples per bay spaced less than a meter apart (81 cm): Deneux's drawings in the archives CRMH, D.6990 and D.6991 show the walls and massive triangular buttresses of Tonnerre. The common rafters are also of the same scantling as the frames with base ties; thus making the roof one of uniform scantling.

Rectangular in plan, the Byloke hospital ward (16 m x 55 m) is divided into eleven bays with ten principal frames spaced 4.85 m center to center, thus creating long bay intervals along the walls, buttressed on the exterior at those points by wall buttresses 1.7 m deep. In contrast to these widely spaced principals, the light common rafters are only 15 cm apart, thus creating the impression of a timber vaulted space (fig. 3.15) not unlike Tonnerre. The unbuttressed section of the wall is a meter thick and only 8 m. high in contrast to the steeply pitched (60°) roof that rises 22 m to its apex. In reference to masonry support conditions, a recent survey of the site has shown clear indication of foundation problems and thus explains the need to correct the outward displacement of the walls with iron ties, inserted in the seventeenth century. Foundation difficulties were probably not uncommon in medieval hospitals, since they were invariably placed in low areas over or next to streams for sanitation reasons. Hence, the Byloke walls lacked long-term stability given the combined geophysical conditions and roof thrust.

In dramatic contrast to our earlier examples, the structural carpentry of the Byloke roof is simultaneously decorative carpentry, as seen in the interior's trefoil design, the molding of the principal timbers, and the impressive scantling of the short principals. In fact, the profile of these great halved timbers is so important aesthetically that the carpenter-designer occasionally pieced in additional wood to achieve the desired curvature (fig. 3.16). The carpenter also ambitiously created a torus molding on the inner face of each principal to emphasize the continuous profile of the great semicircular arch (fig. 3.15). Whether carved out of the principal itself or applied with pegs as required, this round molding significantly anticipates the considerable decorative expansion of moldings in open roofs of the fourteenth and fifteenth centuries (e.g. Westminster Hall). Moreover, the common rafters between the principal frames with their thin bent ashlar pieces and curved braces were originally covered (at least in the sanctuary area) with paneling applied with iron nails, a fragment of which remains in the easternmost bay of the hall where an altar was located. Hence, the decorative conception of the Byloke carpentry combines curved short principals, molded profiles with the form of a timber-paneled vault. To create the great trefoil contour, the lower principals and intermediate common rafters form a great arch that projects inward from the wall head (fig. 3.16); above the crown of this arch, mortised into the horizontal collar beam, the central tunnel vault is carried on double collar plates and obscures the uppermost framing. This configuration creates in effect a *charpente lambrisé*, or paneled vault, so popular in continental halls of the Middle Ages. The roof's grand design also harmonizes with the original Gothic window tracery (partially visible on the north side towards the west), thus suggesting the collaboration of mason and carpenter as designers.

Figure 3.16: Byloke Hospital, Ghent, Roof: Detail of the lower portion of the short principal pieced at its base to complete the desired profile. Here the torus molding is carved from the section. Also visible are the light-sectioned intermediate frames that oversail the wall. (Photo, author)

CONCLUSION

The monumental thirteenth-century carpentry of the Byloke infirmary hall again raises the question of whether medieval carpenters, unable to calculate forces in their constructions and traditionally using green timber with sapwood were compelled to employ 'uneconomical scantling'. The evidence presented in this essay, however, indicates that carpenters made careful choices in their use of woodland resources, and that they carefully adjusted timber scantling to function, both decorative and structural. In fact, this appears to be the case even in grand, high status buildings where economic limitations may not have been so critical as in other instances. Like the nave of Notre-Dame, the seemingly massive roof of Byloke, with its hundreds of members, reveals that really big timbers were used with considerable discretion, especially since the timber had to be imported from the Ardennes forest.[45] Moreover, the Byloke roof's slender common rafters (c. 18 cm in section and about 30 m long) are generally scarf-jointed near the mid-point to avoid using excessively long timbers, thus again utilizing Rackham's Class II size oak of less than 28 cm in girth—a timber section that indeed predominates as noted at Notre-Dame and Cressing Temple. We may conclude, therefore, that the pioneering technical achievement of thirteenth-century carpenters was their economy of construction, the exploitation of slender tension hangers, and especially, the invention of multi-tiered construction covering vast spans while using timbers of moderate scantling to achieve roofs of monumental grandeur.

---

[45] Hoffsummer, *Les charpentes*: 'La centaine (!) de chevrons-fermes justifié sans doute l'importation des bois dont l'origine, d'apès la dendrochronologie, pourrait bien se situer dans le basin de la Meuse'., p. 88. [Some one-hundred samples of rafter couples establishes without a doubt that the imported wood, according to the dendrochronological evidence, came from the Meuse basin].

4

The Gothic Barn of England:
Icon of Prestige and Authority

Niall Brady

INTRODUCTION

The medieval barn is by far the most impressive agricultural building to survive from that period. A good deal of history remains hidden in and around these structures, though until now, students of technology and history have usually left the the study of barns to archaeologists and architectural historians.[1] Their researches have contributed wonderful information about vernacular architecture, but historians have tended to note these buildings only in passing.[2] Not trained in material culture, they are not generally interested in studying building documents, despite recent pleas for someone to look more directly at barns.[3] This essay will try to respond by investigating the motivation behind barn-building in late medieval England.

I thank Paul Hyams and Christian R. Jensen for their comments on a draft of this paper.

[1] There is a considerable bibliography on barns that has focused on appreciating the architectural style of the buildings. Much of the work has dealt with small groups and individual sites; there has been little in way of large regional studies, let alone a national brief. Among the more important works are Walter Horn and Ernest Born, *The Barns of Beaulieu at its Grange at Great Coxwell & Beaulieu-St Leonards*, Berkeley, 1965; S.E. Rigold, 'Some Major Kentish Barns', *Archaeologia Cantiana* 81 (1966),1-30; F.W.B. Charles and Walter Horn, 'The Cruck-Built Barn of Frocester Court Farm, Gloucestershire, England', *Journal of the Society of Architectural Historians* 42 (1983) 211-37; John Weller, *Grangia & Orreum. The Medieval Barn: A Nomenclature*, Suffolk, 1986; C. James Bond and John Weller, 'The Somerset Barns of Glastonbury Abbey', in *The Archaeology and History of Glastonbury Abbey. Essays in Honour of the Ninetieth Birthday of C.A. Ralegh Radford*, eds. Lesley Abrams and James Carley, Suffolk and Rochester, 1991, 57-87; Dave Stenning, 'The Cressing Barns and the Early Development of Barns in South-East England', in *Cressing Temple. A Templar and Hospitaller Manor in Essex*, ed. D.D. Andrews, Chelmsford, 1993, 51-75.

[2] M. M. Postan, *The Medieval Economy and Society*, Harmondsworth, 1972, 114-5, merely cited the barn and the byre as examples of estate buildings that reflect what he recognized was the small amount of productive investment characteristic of manorial agriculture in England between the late twelfth and the fourteenth centuries.

[3] J.P. Greene, *Medieval Monasteries*, London and New York, 1992, 134-45, while Michael Aston, *Monasteries*, London, 1993, 160-61, provides a bibliography of the coverage to date.

Lots of big, magnificent barns survive from the late medieval period.[4] People today often wrongly call them Tithe Barns. Dedicated tithe barns (*grangia ad decimas*) did exist, but they were a special and most probably smaller structure, set apart from the main barn(s).[5] The term had a specific and limited medieval usage, but it has subsequently been applied to all barns regardless of type. The earliest standing remains of the big barns that interest me belong to the late twelfth century.[6] At this time, there supposedly occurred a sharp break in the system of land management that resulted in the birth of 'high farming'—a term borrowed from historians of nineteenth-century Germany which medievalists now use to describe features particular to England.[7] Most lords throughout Europe leased their estates to others, though from c. 1170 into the fourteenth century, English lords assumed management of a section of their lands (specifically the demesne). To this change we owe much of our best evidence for thirteenth- and fourteenth-century estate management, which includes annual accounts to control input and output, court rolls, surveys and inventories, as well as treatises which helped train

---

[4] My research is based on almost three hundred sites identified in a nine-county study area, stretching across the country from Essex and Kent in the east, through Hertfordshire, Middlesex, Oxfordshire, Berkshire, Wiltshire, Gloucestershire, and on to Somerset in the west. A detailed inventory of these sites is included in my doctoral dissertation, 'The Sacred Barn. Barn-Building in Southern England, 1100-1550: A Study of Grain Storage Technology and Its Cultural Context', PhD Cornell University, 1996, unpublished.

[5] The property leases for St Paul's Cathedral, London, make this clear. At Barling, Essex, a *grangia decimae* was recorded in 1281 in addition to a large barn, while in a subsequent record of 1335 the tithe barn was said to be situated close to the outer gate. At Tillingham, Essex, a *grangiam ad decimas* was one of three barns in an outer courtyard, while at Drayton, Middlesex, there was a building *ad reponendum decimas ville*; Guildhall Library 25122/684, 1112; GL 25122/1329; GL Liber I, 25516 f.174v. I am grateful to John Blair for transcriptions of these documents.

[6] The Cistercian barn at Grange Farm, Coggeshall, Essex, is thought to date to the late twelfth century. Like many timber barns, it reveals a complex constructional history and very little of the original fabric remains. There is enough, however, to infer that it was originally quite large, and not dissimilar to the current dimensions of 36.6 m in length, 1.6 m in width, and 10.3 m in greatest height. See Cecil Hewett, *English Historical Carpentry*, Chichester, 1980, 47-9; D.D. Andrews and J. Boutwood, 'Coggeshall Barn. Notes on Discoveries Made During the 1983-84 Restoration', *Essex Archaeology and History* 16 (1984) 150-53; Stenning 'Cressing Barns', pp. 54-6

[7] For an overview, see Edward Miller and John Hatcher, *Medieval England: Rural Society and Economic Change 1086-1348*, London and New York, 1978, 213-39. Since the bulk of the evidence comes from medium-sized and large estates, most of which were run by ecclesiastical lords, it is impossible to know whether direct management, and thus 'high farming', ever became more common than leasing; T. Evans and Rosamund Faith, 'College Estates and University Finances 1350-1500', in *The History of the University of Oxford*, eds. J.I. Catto and Ralph Evans, Oxford, 1992, 671.

clerks on the subject of estate organization and accounting.[8] The accepted view interprets this plethora of information as a real effort at rational economic policy to achieve the highest productivity. Lords thus employed the latest technologies to facilitate large-scale programs of reclamation, construction projects, and experimentation with more intensive techniques of production. They also squeezed the most out of the manorial officials and their villein workers.

Big barns supposedly represented a capital investment aimed at raising productivity. Yet it is hard to see how this could be the case, since their tremendous expense in terms of the annual revenues of a manor undermined such a justification. Did barns raise rates of return? Did English lords have any notion of rates of return anyway? Did their construction reduce the amount of grain that spoiled or was stolen? Certainly the centralizing of large quantities of grain in one location that could be locked-up and guarded must have been a desirable measure. The ability to thresh grain on the spot in convenient circumstances would also have been useful. However, as contemporaries knew, cheaper storage alternatives existed in the form of open-air stacks. A rational economic argument alone remains unconvincing. Other kinds of justification for this level of expenditure lie at least partly in the role barns played as symbols of prestige and authority. We may begin, however, with a brief consideration of their structure and use.

FORM AND FUNCTION

Walter Horn and Ernest Born's publication of the Beaulieu Abbey barn at Great Coxwell, Oxfordshire (formerly Berkshire), shows the magnificence that these buildings can achieve (fig. 4.1). The Cistercians were granted land to found an abbey on the royal manor of Faringdon in 1203, and soon after moved to Beaulieu in the New Forest, but Faringdon and its nearby granges, of which Great Coxwell was one, remained important to the house. In 1269-70, Coxwell contributed almost a quarter of the grain grown on these manors.[9]

---

[8] On the process of account writing, see the introduction to Paul Harvey, ed., *Manorial Records of Cuxham, Oxfordshire circa 1200-1359*, Oxfordshire Record Society 50 (1976) 1-71. On the treatises, see Dorothea Oschinsky, ed. and transl., *Walter of Henley and other Treatises on Estate Management and Accounting*, Oxford, 1971.

[9] S.F. Hockey, ed., *The Account Book of Beaulieu Abbey*, Camden Society Fourth Series 16 (1975) 25.

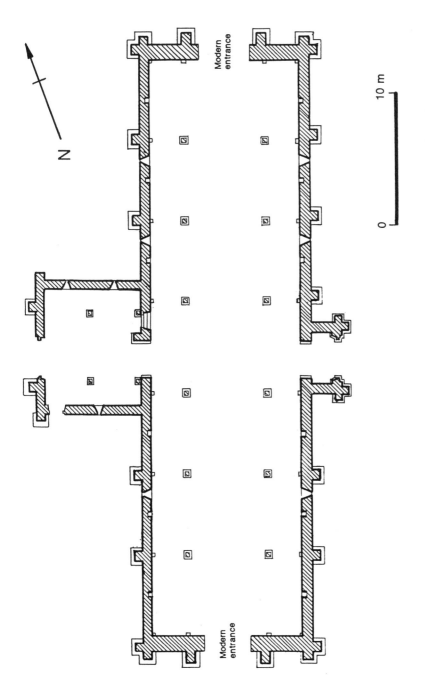

N

10 m

0

Modern
entrance

Modern
entrance

Figure 4.1. Great Coxwell barn, Oxfordshire (formerly Berkshire): plan (after Horn and Born).

The barn that stands today was built during the first decade of the fourteenth century.[10] It is a large rectangular, stone-built structure with massive transeptal entrances in the side-walls and a high, steeply pitched roof covered with slates.[11] As the visitor leaves the lofty buttressed exterior, and passes through the central wagon porch, he walks beneath what was a second floor space, perhaps an office area, and enters a central area over 43 m long and 11 m wide, enclosing a space over 3,800 m³ in size. The eye is drawn along the tall slender arcade posts resting on stone piers, and along the secondary rafters that run from intermediary roof trusses, all of which supports a timber-frame roof that disappears into the darkness over 12 m above ground level. The cathedral-like proportions and layout of this barn are a fine monument to the skilled designers and masons employed in its building. Great Coxwell is perhaps the jewel among the standing medieval barns today. Many others are far less spectacular, but it is true to say that they are all magnificent structures. It has been said that the barn is the largest and most imposing of the manorial work buildings, and the only one to come down to us in considerable numbers.[12]

The primary function of these barns (Latin–*(h)orreum* and *grangia*, Middle English–*bern*) was as a place to store harvested crops.[13] Unlike the barns of North America today, the English barns did not house livestock as well. These buildings were designed specifically with the products of arable husbandry in mind, and it is not uncommon to find them referred to in the sources as the *grangia frumenti*, or the *grangia avenarum*.[14] The distinction

[10] Horn and Born, *Barns of Beaulieu*, p. 35, argue that a date early in the thirteenth century would be appropriate, but a range of timber samples from pads below the aisle posts and from the northern porch were subsequently determined by tree-ring dating to have been felled between 1282 and 1305; John Fletcher, 'A List of Tree-Ring Dates for Building Timber in Southern England and Wales', *Vernacular Architecture* 11 (1980) 34.

[11] It is a distinguishing feature of English barns that the entrances are in the side walls; on the Continent, it is often the case that barns are entered from the end walls.

[12] Jean LePatourel, 'England', *The Agrarian History of England and Wales. III. 1348-1500*, ed. Edward Miller, Cambridge and New York, 1991, 866.

[13] *Orreum* was only ever used to refer to a barn by this period, but *grangia* was used interchangeably to denote a barn or the curial complex of buildings in which the barn would stand. Only the context will reveal which meaning is to be understood; Weller, *Grangia & Orreum*, pp. 28-32. Our knowledge of the agricultural function of these buildings is based on the survival of a number of descriptions of their contents (see n. 15). It must be admitted, however, that these descriptions date from the twelfth to the fourteenth centuries only, leaving open the possibility, however minute, that these barns served other agricultural uses in subsequent years.

[14] A lease of 1174 x 80 between the Canons of St Paul's and Richard Ruffus, canon, for their property at Belchamp St Paul's, Essex, lists a *grangiam frumentariam* and a *grangiam avenariam*; William Hale, ed., *The Domesday of St Paul's of the Year M.CC.XXII; or*

survives on sites even today; on the former Knights Templar's preceptory at Cressing Temple in Essex, the two thirteenth-century timber barns are known as the Barley and Wheat Barn respectively. Even hay was characteristically stored in its own separate building, the *domus feni* or *grangia feni*, while livestock were accommodated in cow houses and sheepfolds, *vaccaria* and *bercaria* respectively. Several early leases from the period will further refer to the crops stored within a barn, since the lessee was expected to return this amount at the end of the lease.[15] There is never any mention of items other than crops, and while there is always the possibility that leases are somewhat idealized in the sense that they record what should happen, rather than what actually happened, the presence of other buildings specifically devoted to hay, stock, and equipment, suggests that the practice may not have been altogether different from the theory, at least on the larger complexes.

The barn was only a temporary storage point for the crops. It provided a large enclosed space in which the harvest could complete its drying process. The crops were bundled into sheaves as they were being harvested, which were then made into heaps within the barn. Over the winter months, they would be threshed and winnowed to separate the ears of grain from their stalks. The entrance bays present an ideal and convenient place for such tasks. When the doors are thrown open, the airflow across the bays can create a draught strong enough to separate the chaff of the stalks from the grain. These bays often have a rammed clay floor, but they might equally consist of a special threshing floor of corrugated stonework, as survives at Shaftesbury Abbey's Barton Farm manor in Bradford-on-Avon.[16] Teams of supervised men would systematically beat the sheaves with flails. Once separated, the grain was moved to the granary. This was sometimes a small area of the much larger barn, but it was typically a different building altogether.[17] So ended the storage function of the barn, or at least what we know of it.

---

registrum de Visitatione Maneriorum per Robertum Decanum, Camden Society 69 (1858) pp. 138-9. In medieval usage, *frumentum* has two translations: in a context where other grains are noted by name (*avenae*–oats, *ordium/ordeum*–barley, *siligo*–rye) it means wheat, but in contexts where the types of grain are not specified, it appears to refer to grain in general, or corn. Corn in this sense is not to be confused with the modern American usage.

[15] At Belchamp, the east side of the wheat barn was filled with wheat behind the entrance, the west with maslin (a rye and wheat mixture). The north aisle of the west end should have rye, and the west half including the south aisle should have wheat. The oats barn was similarly divided in two. The west side was to be filled with oats, the east end-aisle with oats and barley up to the tie beam, and its south side-aisle with barley. The remainder of the barn was to be empty; ibid., pp. 138-9.

[16] *The Medieval Tithe Barn. Bradford-on-Avon, Wiltshire*, Department of the Environment Pamphlet, reprinted 1972.

[17] A fine stone-built granary survives beside the great fourteenth-century barn in Bradford-on-Avon. It was originally constructed on raised piers, which have now been filled in.

Little has been written on the development of barn design in the late medieval period. Study of these buildings has focused on questions of style rather than function, and we remain quite ignorant about the progress of such basic but important matters as ventilation and the means by which these buildings were designed to expedite loading.[18] It is far beyond the scope of the present essay to work through this material, and I plan to examine it on another occasion. I will also refrain from commenting on chronology because the number of narrowly-dated barns is at present rather small.[19] The purpose of my present endeavor is instead to show how humble agricultural buildings can be studied in a new and innovative way. Big barns are impressive structures in their own right, but they are also important for furthering our understanding of social status and the perception of labor in the Middle Ages.

THE ECONOMICS OF BARN-BUILDING

In their attention to detail, clerks drawing up the annual accounts for a manor often recorded the various building enterprises down to the last nail. It is not uncommon to discover a near-complete record describing the building of a barn, and this can provide an exceptionally useful means of seeing its building within the context of the manor's economy. I will highlight the costs involved by looking at a number of barn-building campaigns, and in particular at that carried out by New College, Oxford, at Swalcliffe rectory in the early 1400s. We will then consider the capacity of these buildings to store large amounts of grain. The justification for their construction is not easily explained in terms of the quasi-capitalist enterprises we have grown accustomed to expect of the period. Today we usually expect a return on our investments, and it is not unrealistic to expect this to have been the case in the Middle Ages as well. I will consider what other factors might have influenced their building in the final section of the paper.

We may begin by looking at the costs involved in building a typical large barn. Oliver Rackham has extracted the necessary details for one at Gamlingay, Cambridgeshire, a manor of Merton College, Oxford.[20] The barn no longer survives, but it was timber-built, possibly employing a five-bay design and similar in size to the Templar barns at Cressing, Essex, which are

---

[18] Such problems are wrestled with in Weller, *Grangia & Orreum*, while Bond and Weller end their paper fully aware of the need for useful analysis in this area. The greater project on which their paper is based promises to take the subject further; 'Somerset Barns', pp. 84-7.

[19] Of the three hundred-odd standing and now-destroyed sites described in my doctoral dissertation, dendrochronological determinations are available for only twelve of these barns.

[20] Oliver Rackham, 'Woodland Management and Timber Economy as Evidenced by the Buildings at Cressing Temple', in *Cressing Temple*, ed. Andrews, pp. 90-1, citing Merton College Records 5409, 5410.

rather large structures.[21] The Gamlingay barn was built during 1358 and 1359, and the college was able to cut costs almost in half by using timber from its own woods. Piece-workers were employed for much of the labor, and there was some payment in kind. The master carpenter, Geoffrey Silvester, received almost one fifth of the total value of the contract. He was paid £6 13s. 4d. along with the two quarters of grain in 1358, and in the following year he was given a further 2s. for supervisory visits to the site, and 13s. 4d. for his gown. The laborers were on a daily wage of 3d., and their tasks ranged from felling and preparing the timber to erecting the barn. Simon Selkman was allotted thirty cartloads of firewood in 1358 to supplement a cash payment of £17 for seventy-four thousand roof tiles, while two men employed to apply daub to the walls the following year were paid four bushels of wheat in addition to their wages of 40s.. In all, Merton spent £40 in cash to build its barn—assembling and transporting the necessary materials, and in wages. This is not an inconsiderable sum, and it must have represented a sizable investment for the college. The annual revenues for the manor were around £50. The barn thus required just under half the yearly earnings for two years in a row.

The continuous and detailed accounts of the New College rectory at Swalcliffe, Oxfordshire, provide an opportunity to study the pattern of investment in greater detail, and over several decades. The rectory was presented to the college after William of Wykeham acquired it in 1389. It consisted of tithes, money payments, and a small glebe centered on the rectory but also extending to neighboring Shutford and Epwell.[22] The college leased the rectory, farmed the rectory and the tithes, and received an annual income similar to that of Gamlingay in the order of £50. The accounts mention the construction of a large stone barn in the opening years of the fifteenth century, and it is reasonable to suppose that this was the more magnificent of two fine barns which still stand today.[23] The barn is built mostly from blocks of local Banbury ironstone, with some non-local and rather costly oolitic limestone in the copings and the string courses.[24] Two large entrance porches on the east wall face onto the yard, while smaller porchless doorways lie opposite in the west wall. It is a long barn, divided into ten bays, and measuring 39 m long

---

[21] See below, p. 92.

[22] *Victoria County History, Oxfordshire*, v. 10, ed. Alan Crossley, London, 1972, 237.

[23] R.A. Chambers, A. Fleming, J. Munby, J. Steane, M. Taylor, 'Swalcliffe: A New College Farm in the Fifteenth Century', *Oxoniensia*, forthcoming. I am grateful to Julian Munby for making a draft copy available to me, and to Bronac Holden for providing transcriptions of the Swalcliffe accounts.

[24] In 1405, 368 feet of freestone was purchased for just these areas of the barn at 2d. a foot for a total of £3 18s. 7d.. It was transported some thirty miles from Winchcombe, Gloucestershire (where oolitic limestone is common even today) in eighteen carts costing £1 15s. 10d.—almost half the purchase price; New College Archives (hereafter NCA) 7001/6.

internally, and almost seven meters wide. The tall side walls (just short of 5 m in height) and end walls are elegantly buttressed to support a cruck-framed timber roof that is built in a style typical of the north Oxfordshire area. The roof-ridge is almost 10 m above the ground. The barn forms the western side to a workyard which lies west of the rectory house. The lesser barn lies at right angles to the larger structure on the north side of the yard.

The monies invested in building the barn are indicated in Table 4.2 (see below p 107). This summarizes information for the period between the earliest surviving lease in 1397-99 and 1435-36.[25] Building expenses are generally recorded in the accounts under the headings *custos domorum* and *novum edificium* ('building cost' and 'new building' respectively). It is not uncommon to find details under other rubrics as well, such as the *expense domini custodis et aliorum*—those expenses incurred by the warden of the college and the fellows when they visited the worksite or purchased materials for it. In Table 4.2, I have distinguished between barn-building expenses and those invested in other construction work, be it on other yard buildings (sheep house, stables, granary) or the domestic range (hall, camera, kitchen, chapel). I have also attempted to show these figures in the context of the rectory's overall budget by including the cash receipts, cash expenses, and the assessed worth, or profit, of the rectory, as noted in the accounts.

The receipts (*summa totalis recepte cum arrearagiis*) came from three principal sources: the annual *firma*, or rent, due to the college from the lessee; gifts to the rectory; and arrears owing by the lessee or his predecessors.[26] The rent at Swalcliffe remained fairly constant throughout the period, between £50 and £52 per annum. The level of arrears, however, fluctuated, and this served to inflate the overall receipts total. The large amount of £80 7s. 11d. in 1400-01, for instance, was because the leaseholder, John As, had fallen into arrears totaling £27 6½d. over the three previous years of the lease. The receipts total was thus something of a hypothetical income in the sense that while the monies were due to the college, it was not always paid. The expenses (*summa omnium expensarum et liberacionum*) constituted what the lessee had spent on the college's behalf in the course of the year, and it represents the payment of the rent up to that point. Any difference below the value of the rent had to be

---

[25] A number of overlapping accounts occur. Between 1404 and 1406 separate accounts were submitted by the leaseholder and by the supervisor of building works. The latter constitute a detailed inventory of the sundry building expenses incurred.

[26] The receipts for other manors would also include proceeds from the sale of manorial resources, such as timber, and the proceeds derived from the lord's court. As Swalcliffe had no such resources to sell, and the rectory was not a manorial court, these categories do not occur.

paid over by the lessee, while the college would refund or credit any surplus.[27] In three instances (1404, 1405, 1405-06a), the cash building expenses, which combine the amounts in the Barn and Other Buildings columns in Table 4.2, alone exceed by some small degree the totals stated as the overall expenditure. While this looks wholly anomalous, it was facilitated by certain cash payments to the rectory by the college, which are itemized in the receipts section. £15 16s. 10d., for instance, was paid to the rectory in 1405 to help defray expenses in May and June. The lessees were not responsible for this money, and so there is in fact no contradiction. However, it does raise the prickly issue of how representative the *summa omnium expensarum* was of the total expenditure on the property. This requires a level of detailed analysis of each account that I have not undertaken. Rather, I show the totals only as guides to the cash income and outgoings of the property. The final total included in Table 4.2 was the perceived value, or profit, of the rectory. This was calculated in twelve cases between 1398-99 and 1413-14. The amount is called by various names in accounts of this period, and those used to describe it in the Swalcliffe *compoti* were *valet* and *profectus*.[28] Typical of their type, the accounts do not show the mechanism of its calculation. *Valet* is a figure added after the audit to the foot of the cash account; it was the final word on the matter and served to close the account.

The picture which these figures describe indicates the careful management of the monies accruing to the college from Swalcliffe. Total cash expenses only exceeded the receipts in three instances (1405-06a, 1416-17, 1421-22), and on such occasions only by a matter of a few shillings. The college did not typically spend more on the rectory than it earned from its rent, and it frequently spent close to the entire amount on it; in eight accounts the expenditure and revenue are perfectly balanced.[29] The profit figures, for their part, show the rectory to be a high-yielding source of income for the college. Over the fifteen or sixteen years that it appears, the *valet* or *profectus* was relatively stable at a high level. A slight decline is evident from £53 2s. 9d. to

---

[27] As occurred in 1405-06, when the the vicar, William Ingram, was owed 3s. 2d.. This is where we see something of the dynamics of the medieval account, as described by Paul Harvey, *Manorial Records*, pp. 35-71, where the lessee and the auditors could wrangle over what was owed, and the lessee would often claim compensation for other payments (*allocacione*) made, hoping to reduce the arrears. In 1408-09, John Wylkins, lessee of the rectory, and Thomas Mason, lessee of Epwell, were able to claim additional payments to the sum of 19s. 11d., reducing the amount owed to £11.

[28] Ibid., pp. 56-57; David Postles, 'The Perception of Profit before the Leasing of Demesnes', *Agricultural History Review* 34 (1986) 12-28.

[29] 1403-04, 1404, 1404-05, 1409-10a, 1411-12, 1418-19, 1419-20, 1420-21. The greatest contrast occurred in 1400-01, when the receipts totaled more than £80 and the expenses were less than £54, but as we have noted above (p. 92), this was due to the significant level of arrears that had accumulated.

£49 17s. 11d., with a peak in 1404-05 at £53 9s. 4d. and a low of £41 14s. 2d in 1410-11, but the normal amount of over £50 places the rectory on a par with another equally lucrative rectorial property of the college in Adderbury, which was averaging £53 in these years. This was quite a bit more than other college manors at Alton Barnes, in Wiltshire, Upper Heyford, in Oxfordshire, and Widdington, in Essex, where the *valet* was anything between £27–£35, around £33, and £23 respectively.[30] New College had a large number of these small- to middling-sized manors scattered over Oxfordshire, Wiltshire, Hampshire, Sussex, Essex, and Cambridgeshire, which generated a total annual income of between £500 and £600 in these years.[31] In this context, Swalcliffe was clearly among the more important financial earners for the college.

The specific monies devoted to buildings varied greatly, and could be as little as the total building expenditure recorded for 1435-36 of 5d., when timber purchased for repairs to the granary was trimmed. Over the longer term, however, between 1398-99 and 1435-36, the average expenditure was £6 12s. per annum, or about 14.3 per cent of the mean total expenses recorded for the same period.[32] Fourteen per cent represents a significant level of investment in buildings alone, and there is no avoiding the fact that the lion's share was spent building the barn. Over the thirty-eight year period of the accounts represented up to 1435-36, New College spent almost £290 constructing, altering, and maintaining its buildings at the rectory and at the satellite properties of its glebe in neighboring Epwell and Shutford. The building of the Swalcliffe barn took place over a six-year period between 1400-01 and 1406-07, where it dominates the accounts. Much of the work was done during 1404 and 1405. This construction absorbed a minimum of £123,

---

[30] The following figures are drawn from the account rolls on all four manors for the years up to 1426-26: Adderbury, Oxfordshire: 1394-95 £60 5s. 71/2d., 1398-99 £53 6s. 5d., 1399-1400 £53 12s. 4d., 1400-01 £53 16s. 7d., 1401-02 £52 13s. 5d., 1405-06 £53 8s. 5d., 1408-09 £53 16s. 1d. (NCA 5800/9, 12, 5801/1-3, 7, 4). Alton Barnes, Wiltshire: 1389-90 £15 7s. 8d., 1393-94 £25 12s. 9d., 1394-95 £25 6s., 1395-96 £24 7s., 1396-97 £28 19s. 11d., 1397-98 £30 7s. 10d., 1398-99 £30 6s. 3d., 1399-1400 £27 3s. 8d., 1400-01 £34, 1401-02 £37, 1402-03 £32 8d., 1403-04 £31, 1405-06 £25, 1407-08 £30 2s., 1408-09 £35 10s. 4d., 1410-11 £34, 1411-12 £28 18s. 10d., 1412-13 £31 (NCA 5816, 5820-29, 5831, 5833, 5835, 5838, 5841-43). Upper Heyford, Oxfordshire: 1398-99 £36 8s. 10d., 1399-1400 £34 16s. 31/2d., 1400-01 £34 3s., 1401-02 £31 18s., 1402-03 £32 15s. 2d., 1403-04 £28 9s. 4d., 1404-05 £32 6s., 1406-07 £34 14d., 1407-08 £36 11s. 111/2d., 1408-09 £34 11s. 6d., 1409-10 £30 9s. 3d., 1410-11 £35 2s. 71/2d. 1411-12 £32 18s. (NCA 6298-6301, 6303-05, 6307-6312). Widdington, Essex: 1395-96 £23, 1407-09 £23 11s. 9d. (NCA 7227, 7239).

[31] Evans and Faith, 'College Estates', pp. 691-2.

[32] This does not include the £53 reinvested in the barn project in 1403-04, since presumably this appears in the building expenses figures over the next few years.

while the full figure was somewhat more.[33] This represents a minimum of 42 per cent of all the monies spent on buildings over almost four decades. No other structure at Swalcliffe received so much attention as the barn, and only the domestic range rivalled the quality of materials employed in its construction.

An older barn still stood when the new barn was being built. It is cited in the earliest account of 1397-99, needing repairs to the frame of its roof, as well as some thatching. It remained standing for another decade, when its walls were taken down and broken up in 1409-10. As there is no mention of its roof at this point, it is possible that this had already collapsed and the barn was no longer used to store grain. In 1403-04, just as construction of the new barn was getting fully under way, the project was given a significant injection of capital when the entire profit for the rectory in that year, £53, was given towards the new building; *et profect 'rector' ibidem hoc annum liii Li. et totum in novum edificacione.*[34] This undoubtedly facilitated the expenditure over the next two years when more than £80 was spent between materials and wages. The progress of the building can be traced. The primary roof timbers had already been selected, and preliminary work on the building begun. In 1404, the building season lasted eight months, from early February right through the summer to Michaelmas, 29 September. The site was a hive of activity, and the walls were the focus of attention. £14 6s. 61/2d. was spent on laborers' wages to quarry the stone, and an unknown portion paid to masons to build the walls. In 1405, lime was brought in from Ascot and Ayno,[35] while laths and other roof timber, along with various types of nails were purchased in Stratford-on-Avon and brought to the site. Slates were bought by the thousand, and the oolitic limestone for the copings and string course arrived from Winchcombe. The roof was raised in this year (*oper' levat' meremium dict' grang'*), and there was something of a celebration afterwards, the whole costing £2 7s. 3½d. William Ingram, the vicar, was rewarded with £1 for his expeditious work as supervisor. Wages in 1405, as they had been the previous year, were by far the most costly payment, amounting to £43 14s. 11d. alone. In 1406, the major work had been completed, and it remained to finish the roofing. Fifty-one thousand slates were acquired to help in the task, at 6s. a thousand, which included their fitting. They came from Scloutere, which may be present-day Upper/Lower Slaughter in Gloucestershire, more than nineteen

---

[33] The uncertainty of the full amount is because the 1404 account mixes some of the expenses with those used to build the presbytery at Shutford, and presents a simple subtotal for both at £40 9s. 1d.

[34] NCA 7001/3.

[35] Aynho is in southern Northamptonshire, Chambers et al., 'Swalcliffe', forthcoming.

miles distant.[36] The slates arrived between 4 May and 30 June, and were transported at a cost of £2 3s. 2½d., about half the purchase price. 1406-07 saw the final touches to the structure, with the construction of two doors, which between them accommodated the not inconsiderable weight of eighty-six pounds of iron in their hinges alone.

While the barn was being built, the only other construction work being done was at Shutford presbytery (in 1404-04). Other work at Swalcliffe was minimal; the walls of the cow house were repaired in 1405-05, and a new one was constructed in 1405-06 at a very small cost, along with a belfry for the rectory. The belfry was mostly paid for by a special gift from the college of 6s. 8d. From 1406-07, with the new barn almost completed, the other buildings begin to receive more attention. Most of the expenses go to repairs and extensions, and the college in later years was to spend a lot of time building walls and gateways within the curial complex and around the perimeter. Other new buildings were also made, and the impression that the accounts convey is that the college was prepared to invest time and money in their construction and maintenance, but no building rivalled the barn in extravagance. The most costly other new structure was the sheep house built in 1409-10 for the not insignificant sum of £35 6s. 1½d.; the timber alone cost £10. Lime, sand, water and earth were bought the following year to create mortar for stone walls, which appear to have consisted entirely of local stone, and the roof was thatched, except for the doorway which was slated. There is no indication of how large the building was. A new pigsty in 1434-35 cost £1 2½d., and used stone from neighboring Tadmarton. In 1440-41, the granary and dovecote were renovated: their walls were repaired, and a new roof was constructed for the granary.[37] The carpenter John Wiltshire was hired for eleven days to do this, and forty rafters along with laths and nails had to be purchased in Stratford and transported, costing 18s. 5d.. A slaterer was hired to repair the tiling at £2, while twenty feet of coping stone cost 3d. a foot. Within the domestic range, the alterations and repairs to the hall, cameras, chapel, and kitchen amount to a steady and constant outlay, but one that is in terms of many shillings rather than pounds.

The barn was the only building in the period of these accounts to receive a wholesale reinvestment of the year's profit. Such an event cannot but stand out, and it begs consideration of the type of investment these buildings were deemed worthy of repaying. Indeed, the cost to the college of building a barn in 1390-91 at Alton Barnes, Wiltshire, dug deeply into available resources. The structure was perhaps more modest than that at Swalcliffe,

---

[36] Chambers et al., 'Swalcliffe', forthcoming.
[37] NCA 7003/21.

costing only £38 11s. 2d.[38] However, the manor's revenue (*profectus*) for that year amounted to only £24 8½d., and averaged £28 for the years 1389-90 to 1412-13.[39] In terms of the manor's economic output, this was a much more expensive barn to build. Its feasibility was probably only possible by making up the difference with funds from the college's general resources.[40]

One of the underlying tenets explaining the pattern of agricultural investment in the late medieval period is that the quasi-capitalistic landowners were very conscious of how much they spent on production overheads in areas such as drainage and other reclamation schemes, and buildings.[41] Postan argued that only a very small proportion of their incomes, perhaps some five per cent, was so invested.[42] Accordingly, there would be little room for excess, and everything would be limited to the barest of essentials as an estate strove to reap the greatest profit. This suggests something of a skinflint attitude on the part of the medieval lord to working the land, and we should then expect the fabric of the barns to reflect this bare-bones attitude. However, this thesis does not fit the facts of the barns very well. How long would it be before the barn at Swalcliffe could repay the more than £123 it cost to build? Less than fifty years after its construction, it needed a complete retiling (1454-55, 1455-56), while in 1446-47 its floor had been raised and leveled.

A small selection of other barns retain certain embellishments that provide further opportunity to question the supposed parsimony of medieval landlords. At Great Coxwell the intermediate roof trusses rest on ornate corbel mouldings; at the Canon's Barn, Wells, Somerset, the roof was supported by massive rounded stone piers which rose over five meters from the ground; in the Glastonbury Abbey barns at Glastonbury and Pilton, Somerset, the symbols of the evangelists are carved in roundels below the entrance gables and those of the end walls, and they are part of a larger assemblage of figure-carving on the exterior walls.[43] These have little obvious place in an economically rational system of capital investment in agriculture.

---

[38] NCA 5817. A timber barn stood here until recently. It had clearly been modified in post-medieval times, but the medieval portion may have been a ten-bay construction. This would have rivaled Swalcliffe in size, while the lack of stone in its construction probably explains the significantly lower building cost.

[39] See n. 30 above.

[40] Evans and Faith, 'College Estates', p. 691, have calculated that the annual income from the entire estate rose to between £500 and £600 in the 1390s. They elsewhere note (p. 674) that the net receipts at Alton Barnes for 1390-91 were only £8, but unfortunately they do not show how they deduced this figure. The *profectus* of £28 8½d., however, is stated in the account.

[41] M.M. Postan, 'Investment in Medieval Agriculture', *Journal of Economic History* 27 (1967) 576-87.

[42] Ibid., pp. 578-9.

[43] Horn and Born, *Barns of Beaulieu*, pp. 19-31; Sherwin Bailey *Wells Manor of Canon Grange*, Gloucester, 1985, p.5; Bond and Weller, 'Somerset barns', pp. 68-73, 76-9.

A more focused critique might consider barns in terms of their capacity to store large amounts of grain. It is no easy task to know the amount of a manor's produce that was stored in a particular barn. While it is often possible to reconstruct the size of a manor around the time the barn was built, the number of existing barns is another matter altogether, to say nothing of their sizes. The extent to which a barn served to store the immediate manor's crops, or accommodated crops from elsewhere should be considered as well. This is perhaps more of a problem on large estates with many manors (particularly ones which shared resources), than with smaller endeavors.[44] Bond and Weller see the possibility of examining the relationship between barn capacity and acreage on manors which form discrete parts of Glastonbury Abbey's estate, such as the manors of Doulting and Pilton, and I look forward to the results of their investigations.[45]

One case where these matters seem to be resolved is the Knights Templar's preceptory at Cressing Temple, Essex. Cressing was granted by Queen Matilda to the Templars in 1137. At their dissolution in 1312, the lands passed to the Hospitallers.[46] By this time, the two large timber barns which today occupy much of the north-west corner of the complex were already built.[47] The Barley Barn was raised in the early thirteenth century, the Wheat Barn after 1250. The Barley Barn is slightly shorter, but is wider and a little higher than its architecturally more sophisticated neighbor. The barns measure 36.6 m and 39.7 m in internal length respectively, 13.6 m and 12.2 m in width, 11.7 m and 11 m in height to the ridge (7 m and 6.4 m to the tie beams), and the walls are 3 m and 2.6 m high. The estate was 1,287 acres in size in 1309, of which 1,115 were 'profitable' (*terra lucrabili*). An inventory drawn up in 1313, a year after the order was suppressed, listed 601 acres of arable land.[48] John Hunter has suggested that the gross arable acreage in the demesne may have been as much as 781 acres. When one-third is deducted for fallow land,

---

[44] One might suspect that it is also a consideration with barns on the home farm, where commodities continually came in from outlying manors, but the impression I get is that grain tribute arriving at the abbey or estate center was already threshed, and would have gone directly into granaries for storage, not barns. This certainly appears to be the case in the commercial world of the cornmonger. They would buy quantities of grain at the points of production by the quarter and bushel rather than by the sheaf. Since it was usual to measure threshed grain by the quarter and bushel, and unthreshed grain by the sheaf, it indicates that cornmoners were dealing with threshed commodities, and when their purchases arrived at their storage facilities they went directly into granaries.

[45] Bond and Weller, 'Somerset Barns', p. 86.

[46] Pat Ryan, 'The History of Cressing Temple from the Documentary Sources', in *Cressing Temple*, ed. Andrews, pp. 11-5.

[47] Stenning, 'Cressing Barns', pp. 58-66, 68-74.

[48] Ryan, 'History of Cressing', p. 21.

this becomes 521 acres.[49] In order to calculate the storage capacity from such an area, Hunter employs a figure of eighty sheaves an acre, with each sheaf occupying 2 cubic feet (0.566 m^3). An acre of grain would thus occupy 4.5 m^3.[50] Hunter further accepts that the harvest would have been stored up to the tie beams only, and that the side aisles would have been used to stack grain only in times of bumper crops. Accordingly, his figures for the capacity for each barn (2,638 m^3 for both) are based on a small area of their actual sizes, and he concludes that the two barns at the preceptory could readily accommodate the annual harvest between them; up to 582 acres worth of sheaves in ordinary years, and 863 acres in bumper years.[51]

I would argue that Hunter's estimate of storage capacity is a conservative one that is ultimately quite wrong. There is nothing in the primary sources to indicate that side-aisles were only employed for storage during seasons of bumper crops; indeed, the aisles are the logical place to fill with grain first since the carters would need the central naves to be clear as long as possible to facilitate maneuverability.[52] Stacks were also built to various heights, and it is not unknown for them to reach far beyond the height of the tie beams, up to the apex of the roofs. In 1189 at Nettleton, Wiltshire, there was supposed to be a heap that rose only to below the eaves, *sub severundas*, and in a lease of 1174 x 80, at Belchamp St Paul, Essex, a stack of oats in the oat barn extended to the tie beams, *ad trabem*. However, at the sister property of Walton-on-the-Naze, the north and south naves of the *magnum orreum* were filled with oats and wheat right up to the ridge, *ad festum*, some 10 m high and 4.5 m above the tie beams.[53] The sources thus clearly indicate the ability and practice of stacking these barns to capacity.

Any calculation of the storage capacity of a barn should take this into consideration, and thus should ascertain the maximum storage capacity, which

---

[49] John Hunter, 'The Historic Landscape of Cressing Temple and Its Environs', in *Cressing Temple*, ed. Andrews, p. 34.

[50] Hunter uses the figures advanced by John Weller, *Grangia & Orreum*, pp. 10-11. Weller is concerned with oat and wheat crops only. He bases his findings on information contained in the thirteenth-century treatises of agricultural practice and estate management, and on the results of Titow's work on the Winchester accounts; J.Z. Titow, *Winchester Yields: A Study in Medieval Agricultural Productivity*, Cambridge, 1972. Weller's is still the only attempt to estimate sheaf size. It would be instructive at some future point to test the validity of his findings. Oat and wheat sheaves appear to have been small compared to those of rye. While a sheaf length of 80 cm may have been typical, Greig points out that rye straw can sometimes be more than 2 m in length; James Greig, 'Plant Resources', *The Countryside of Medieval England*, eds. Grenville Astill and Annie Grant, Oxford, 1988, 112.

[51] Hunter, 'Historic Landscape', p. 34.

[52] See n. 15 above for how the barns at Belchamp St. Paul, Essex, were loaded.

[53] J. E. Jackson, ed., *Liber Henrici de Soliaco Abbatis Glaston.* [1189] Roxburghe Club, London, 1882, p. 106; Hale, *Domesday of St. Paul's*, pp. 138-9, 129-31.

is easily realized. Once the internal length, width, height-to-ridge, and height-to-eaves-level measurements are known, simply calculate the area of the elongated rectangle formed below eaves level, and add the result to half of a similar sum for the roof space above the eaves. The half value accommodates the triangular space created by the roof.[54] Accordingly, I estimate the storage capacity of the Cressing barns to be 6,852 m³, or almost three times Hunter's figure. If we subtract one-third of this for unused spaces (including the entrance bays) it leaves a total capacity of 2,388 m³ for the Barley Barn and 2,179 m³ for the Wheat Barn. The Barley Barn alone could just about store the potential 2,344 m³ from an eighty-sheaf-an-acre harvest of the preceptory's demesne arable. The two barns together could hold such a harvest from 1,015 acres, which is almost twice the arable acreage on the entire estate. In other words, the barns at Temple Cressing were probably never filled to capacity, and were at best only half-filled after the Wheat Barn was built. This is surely not the product of a niggardly mind, nor could it be an oversight of enthusiastic barn-builders.

The storage needs of the manor of Harmondsworth, Middlesex, can also be estimated. The manor was bought in 1391 by William of Wykeham from the alien abbey of the Holy Trinity (later St. Catherine's), Rouen, and he granted Harmondsworth to Winchester College school, Hampshire. Mention of barns goes back to the twelfth century, though tree-ring dating strongly argues that the current massive timber barn was a new barn built by the school in 1426-27.[55] The barn stretches for twelve bays and has three entrance bays. It is over 58 m long internally, 11 m wide, it extends to almost 12 m high at the roof ridge, and is just over 4 m high at eaves' level. Arable agriculture was the mainstay of the manor.[56] In 1293, the cultivated area consisted of 240 acres of demesne and 233 acres of non-demesne land, and though this oscillated over the next century and a half, it appears to have remained a fairly constant figure. Throughout the fourteenth century and into the fifteenth, almost half the demesne land was given over to wheat, while maslin and barley shared much of the rest, with some peas. Assuming that the barn might have accommodated the crops from across the manor (and it is more than likely that tenants would have their own storage facilities), and cutting out one-third for fallow, leaves an annual crop acreage of 315 acres. The storage requirement for an average yield would be roughly 1,417 m³. The barn at Harmondsworth

---

[54] Bond and Weller, 'Somerset Barns', p. 83, have suggested a two-dimensional sizing scale for barns based on overall length and breadth. The attraction of the method I propose is that it is three-dimensional.

[55] *Victoria County History, Middlesex* 4, eds. J.S. Cockburn and T.F.T. Baker, London, 1971, 7; David Pearce, 'Clues Written in Wood', *Country Life* 184.52 (1990) 40-42.

[56] *VCH Middlesex*, 4, pp. 10-11.

has a volume of 5,288 m^3 which amounts to 3,525 m^3 when one-third is subtracted to allow for unused spaces. This still leaves a barn that is almost two-and-a-half times larger than was needed to store the crops from the entire manor. Sites like Cressing Temple and Harmondsworth serve to call attention to a certain extravagance in barn-building.

Not every barn was necessarily inflated beyond its optimal storage capacity. Great Coxwell, for example, was apparently built to accommodate the size of its harvest. The Faringdon estate, of which Great Coxwell is a part, has not been studied much. I can only make a crude calculation of the storage needs of the barn, based on the corn production figures in the 1269-70 account.[57] The barn was probably built during the first decade of the fourteenth century, so there is a time differential to bear in mind. I have combined all the crop types which the accounts list (wheat, barley, oats, rye, and beans) and assumed the storage requirements relevant to Weller's calculation for oats and wheat. The total figure is 594 quarters, which translates to a possible 38,016 sheaves. This would require 2,138 m^3 of storage space. The barn has a capacity of 3,834 m^3, which reduces to 2,556 m^3 when a third is subtracted for unused spaces. The fourteenth-century barn would then be able to accommodate the 1269-70 harvest, but there would not be any significant amount of excess space left over.

If lords indeed watched their pennies with respect to grain storage, they could have found satisfaction in the far cheaper and quite practical method of outdoor storage in ricks. The evidence for open-air storage is clear. It is also clear that such ricks were not widely employed. In its advice for estimating the expected yield for each stack of grain after threshing, the *Husbandry* treatise recommends that the overseer tally each one separately, '[a]nd if there are stacks outside [the barn] then he ought to have them measured by foot or by rod...'.[58] The *Rules* follows *Husbandry's* approach and advises lords to use loyal men, '...to estimate every year at Michaelmas all the stacks, within and without the barn, of each kind of corn ...'.[59] In practice, it appears that leguminous crops were more likely to be stored in the open than grain. The 1301-2 Winchester Pipe Roll refers to 4d. being paid to cover a stack of beans at Rimpton, Somerset, and stacks of peas and vetches were covered at Downton, Hampshire.[60] However, the twelfth-century lease agreement for the St Paul's manor of Walton-on-the-Naze, Essex, notes the

---

[57] Hockey, *Account Book*, p. 25.

[58] 'E si il yath tas dehors, si le face mesurer par peez ou par rodes...', Oschinsky, *Walter of Henley*, pp. 420-1.

[59] '... a la saint Michel...aesmer trestuz les tas dedenz graunge e dehors de checune manere de ble ...'. Oschinsky, *Walter of Henley*, pp. 392-3.

[60] Hampshire Record Office 11M59/B1/58. My thanks to Mark Page for bringing these references to my attention.

stacking of wheat in the open air as one of the four stacks built up in the *curia* of the manor, and a similar entry for Wrington, Somerset, in 1189, adds that a stack outside the barn should accommodate three cartloads worth of material and be constructed in the expected (and presumably time-honored) manner.[61] It is probably safe to think that such a stack constituted the product of one and more acres.

Unlike a barn, the open-air rick could not have cost any amount of money since there is no expensive structure to build. Among the biological advantages to storage in the open is resistance to certain grain infestations that run riot in a barn. Referring to grain storage in the post medieval period, François Sigaut has noted that '...the European weevil, *sitophilus granarius*, the main insect pest in former times, cannot survive European winters outside granaries or barns—whereas in well-built ricks, being set on a new plot, the grain was not damaged by weevils and less by rodents than a layman might expect'.[62] The major disadvantage of the open-air rick is that once it is opened it has to be fully emptied lest the grain be ruined. This would require immediate threshing of the exposed sheaves and the availability of the labor to do so. On English manors prior to the Black Death, however, such labor would not have been hard to find. Where it was available, we must indeed question the advantages of the barn over the stacks.

What of the added security that a locked-up barn provided? This presents a clear advantage over open-air storage, yet is it too unreasonable to suppose that a secure enclosure built around an area of ricks would provide all the security necessary? Such an arrangement should also more readily prevent the wholesale destruction of the stored crops in the event of a fire, whether resulting from deliberate arson or, more likely, accident.[63]

Once the final threshing was completed, the barns would have been empty and standing idle for an indeterminate length of time, since the sources

---

[61] Hale, *Domesday of St. Paul's*, 130; Wrington: *Extra grangiam debet esse unum tas trium carrearum. modo est tantum quantum debet esse*, Robert De Zouche Hall, 'A Twelfth-Century Barn of Glastonbury Abbey', *Notes and Queries for Somerset and Dorset* 29 (1968) 139.

[62] François Sigaut, 'A Method for Identifying Grain Storage Techniques and Its Application for European Agricultural History', *Tools and Tillage* 6 (1988) 16.

[63] Walter of Henley refers to the threat of fire when he recommends that any surplus grain left over at the end of the agricultural year be converted to cash rather than hoarded, for if there was a fire at least the coin would not be destroyed; Oschinsky, *Walter of Henley*, pp. 309-10. In her study of fourteenth-century crime, Hanawalt notes that arson was not a major crime, accounting for only 0.8 per cent of all crimes recorded. However, it occurred in only one of two main places: the victim's home (62.5 per cent), or their barns (37.5 per cent). It was also by far the most common crime perpetrated in a barn; Barbara Hanawalt, *Crime and Conflict in English Communities 1300-1348*, Cambridge, MA and London, 1979, 77, 90-2.

do not indicate when the threshing ended. Threshing and winnowing were winter tasks undertaken under shelter, which meant that the weather did not directly dictate the speed of the operation. There were likely still sheaves to thresh in the spring, so it was not until summer that the barns were almost fully emptied. If so, this still left two-to-three months of idleness before the new harvest. It is hard to imagine that a society singularly driven by an economically rational spirit of enterprise would make the type of investment that barns represent. So why were they built?

THE BARN AS SYMBOL OF PRESTIGE
The primary storage purpose of a barn does not preclude other functions related to their costly and imposing character. In recent years, researchers have begun to change their understanding of the rationale behind the largesse and ostentatious lifestyle of the secular aristocracy and ecclesiastical landowners. Appalled by the extravagance of their household expenditure, Postan argued that very little of the income from a property was actually saved, but was instead 'squandered' in a variety of ways.[64] In his view, medieval agricultural technology suffered from an endemic lack of productive investment wherein medieval lords, rather than try to improve productivity, instead opted to exploit manual labor by devoting most of their profits to '...unproductive uses: buildings to live in and to pray in, ironmongery and horses to fight with, parks to hunt in, and manpower to serve and to follow'.[65] Unlike this dour assessment of our forebears, Dyer's recent survey of living standards in the period tries to consider medieval economic behavior within the context of the social morality of the time.[66] Instead of castigating the aristocracy for wastefulness and a lack of clear-sightedness in business matters, he draws on the growing body of research that recognizes their astuteness in these areas, observing that their ostentatious lifestyle helped to maintain prevailing social distinctions.[67] Wealthy landowners, as he remarks, were expected to live up to their incomes; heavy expenditure on the household, liveries, and entertainment, was a form of productive investment in its own right because it served to gain respect and honor within society's upper echelons, which in turn paid dividends in the form of royal or other suitable patronage, marriage and other contracts. Wealth and display, in other words were profitable; they helped to create status, and status brought with it power and authority.

---

[64] Postan, 'Investment', pp. 579-80.
[65] Ibid., p. 581.
[66] Christopher Dyer, *Standards of Living in the Later Middle Ages. Social Change in England c. 1200-1520*, Cambridge and New York,1989, 27-108.
[67] Ibid., pp. 89-91. See also Howard Kaminsky's intuitive study, 'Estate, Nobility, and the Exhibition of Estate in the Later Middle Ages', *Speculum* 68 (1993) 684-709.

Prestige as an element in barn-building finds a certain precedent in other building projects of the period.[68] The inflated sizes of the thirteenth-century Hospitaller barns at Cressing Temple and the fifteenth-century one at Harmondsworth tend to take the barn out of the simple role of 'productive investment' envisaged by Postan. Instead, we need to ask to whom was this symbol of show directed and why. There are doubtless a range of audiences and rationales. In the case of the barn at Swalcliffe, with its elegant trim of expensive stonework, it would be quite a suitable emblem of the important college in the countryside as well as a reflection of the self-confidence of the college's founder William of Wykeham, 'sometime bishop of Winchester'.[69] He was quite an entrepreneur in his day, as well as a reformer who set up free colleges in Oxford and Winchester, and possessed an astute ability to endow them with lands that had been perhaps a little run-down, but whose great potential the colleges quickly realized.[70] Large-scale building projects were undertaken throughout the estate in the years following the college's foundation in 1379, and some fine builders were employed, among them the renowned mason, Richard of Winchcombe.[71] The crowning achievement of this very able builder's career was his work on the Divinity School in Oxford (now part of the Bodleian Library), on which he worked during the last decade of his life (1430-40). His delicate touch is seen on buildings throughout Oxfordshire. Not all were New College properties by any means, but the college did employ him at Swalcliffe in 1405-6, where he was paid 2s. to build in stone. He left his distinctive mark at Adderbury, where the college built a new chancel here between 1408-9 and 1417-8 costing almost £400. As primary mason, Winchcombe was personally noted for building the traceried windows, whose light mouldings reflect his recognizable style.[72]

---

[68] Hilary Turner, for instance, suggests that civic pride was among the motivations behind building late medieval town walls. Walls came to express the legal separation of town from countryside, and they also became one of the expected hallmarks of a great city. Accordingly, the magnificent fortifications of Edward I in Wales served to impress the Welsh with the might of English power and prestige; Hilary Turner, *Town Defences in England and Wales. An Architectural and Documentary Study AD 900-1500*, London, 1971, 90-94.

[69] George Moberly, *The Life af William of Wykeham Sometime Bishop of Winchester, and Lord Chancellor of England*, London, 1887.

[70] Wykeham was especially innovative and persistent in his efforts to secure lands of the alien priories. Evans and Faith, 'College Estates', pp. 642-50.

[71] Ibid., p. 678; on Winchcombe's career, see T.F. Hobson, *Adderbury 'Rectoria.' The Manor at Adderbury Belonging to New College, Oxford: The Building of the Chancel 1408-1418: Account Rolls, Deeds and Court Rolls*, Oxfordshire Record Society 8 (1926), pp. 26-41, and John Harvey, *English Mediaeval Architects. A Biographical Dictionary Down to 1550*, London, 1954, 296.

[72] Hobson, *Adderbury 'Rectoria'*, pp. 34-41.

THE BARN AS SYMBOL OF AUTHORITY
Such display also conveyed a message of authority. Just as a Welshman in the time of Edward I undoubtedly felt awed by the sight of the over-built English castles in his home country, so a magnificent barn provided an ideal instrument by which a lord could impress his authority on a distant manor at harvest time. The six weeks between Michaelmas and Martinmas (29 September—11 November) were a particularly busy time of the year, when every tenant and peasant would need to reap and store the product of their own plots, and prepare their lands for the future sowing. Yet the lords claimed a right through customary services to have their crops harvested first. While the common law came to protect tenant interests on the free tenements of a manor, throughout the thirteenth century lords generally dealt with their demesne lands as they saw fit.[73] Demesne properties consequently possessed customary services that provided much of the labor a lord needed. At Ramsey Abbey, each customary tenant in the mid-thirteenth century appears to have owed five half-days of work a week during harvest time where they had to present two men for the work, and three full boon-works (full days) on days of the lord's choosing when they had to bring the entire family.[74] Needless to say, such labor services were not welcomed.[75]

The need to control the workforce was ever-present. According to Walter of Henley, who wrote the most popular treatise on agriculture and estate management of the period, a lord who actively managed his estate would more likely see his workers work for him than against him. A physical presence was considered important, but rarely do the sources permit us to see lords actually engaged in the day-to-day work of agriculture. Ernulf de

---

[73] Paul Hyams, *Kings, Lords, and Peasants in Medieval England. The Common Law of Villeinage in the Twelfth and Thirteenth Centuries*, Oxford, 1980, 52-5. On the inability of Westminster Abbey to keep its customary tenants in line, see Barbara Harvey, *Westminster Abbey and its Estates in the Middle Ages*, Oxford, 1977, 121-2.

[74] Warren O. Ault, 'Open-Field Husbandry and the Village Community', *Transactions of the American Philosophical Society* 55 (1965) 13 n .9.

[75] Hilton argued some time ago that labor services were increased, even doubled in this period; Rodney Hilton, 'Peasant Movements in England before 1381', *Economic History Review* 2 (1949-50) 122-4. He agreed with Postan, who proposed that lords were inclined to impose additional services when and where the demand grew; M. M. Postan, 'The Chronology of Labour Services', *Transactions of the Royal Historical Society* 20 (1937) 174, 187-9. Paul Harvey, however, adopts a less excited tone by arguing that in the restoration of services during the thirteenth century, lords did not invent new burdens for their tenants, but brought back old ones. He makes a case for the unattractiveness of labor services for the lords, but they were pursued because an inflationary climate in the period made commutation of these services a less than realistic alternative; Paul Harvey, 'The English Inflation of 1180-1220', *Past and Present* 61 (1973) 22.

Hesdin, lord of Chipping Norton, Oxford, is an exception. In praising Ernulf's piety, William of Malmesbury tells of how this great and successful twelfth-century lord would stand at the entrance to his barn at harvest time to make sure that the proper amount of tithe was taken from the loads of sheaves as they came in from the fields.[76] References to the actions of the lord's officers are more common in this regard. Walter encourages the presence of the manorial bailiffs and reeves at the first day of fallowing, stirring, and sowing to ascertain how much work was achieved, and to use this as the yardstick for the rest of such labors. Presumably, the enthusiasm to work would be greatest on the first days, but would decline rapidly as time went by. This initial measure provided a means of control. Implementing it was another matter. Since the laborers, '...customarily doe loyter in theire woorke...', the officials were also encouraged to '...lye in wayte against theire frawde'. The image of bailiffs hiding behind hedges would be almost comical if it was not taken so seriously.[77]

These same officials were not above suspicion either. A 'faythful man' formed a necessary adjunct at the barn by independently recording the issues of grain, '...for often tymes a man shal perceave that the barnekeeper and the garneter doe ioyne together to doe falslye'.[78] Another treatise identified such a man as a member of the household staff.[79] Surviving accounts evince similar concerns by showing little of the interest in ascertaining the *profectus/valor* of a manor expected of a profit-conscious world.[80] Paul Harvey has likened the use of accounts to calculate profit to that of the modern check-book as a means of ascertaining a person's income; '...a possible exercise, but not one

---

[76] *Willielmi Malmesbirensis monachi de gestis pontificium anglorum*, ed. N.E.S.A. Hamilton, Rolls Series, London, 1870, 437-8; discussed by Reginald Lennard, *Rural England 1086-1135*, Oxford, 1959, 69, 210-12.

[77] Oschinsky, *Walter of Henley*, pp. 316-7. Ault's study of the village by-laws focuses more on petty crime (theft and trespass) than on idleness, but it shows the village equivalent of the lord's bailiff in action against various transgressors. At Newton Longeville, Buckinghamshire, for instance, two men were caught taking grain from the field 'contrary to ordinance' in 1329 and were fined 2d. each, while three men had handed over sheaves in the field, and were fined 6d. There was also a whole slew of ordinances against carting grain at night; Ault, 'Open-Field Husbandry,' pp. 4-7, 16-9. One can well imagine the wardens being abroad in the dark hours after a harvest day, patrolling for artful dodgers.

[78] Oschinsky, *Walter of Henley*, pp. 322-3.

[79] Ibid., pp. 394-5.

[80] The *profectus* calculation was a feature of the thirteenth and early fourteenth centuries, and though it was sometimes an involved task, and may have helped landowners decide whether to maintain direct management or lease their demesnes, it did not become a standard accounting procedure. Its most popular employment appears to have been between ca. 1290 and 1320; E. Stone, 'Profit and Loss Accountancy at Norwich Cathedral Priory', *Transactions of the Royal Historical Society* 12 (1962) 25-48; Postles, 'Perception of Profit'.

for which the form of the document was intended...'.[81] The honesty of the manorial officers in their handling of their lord's property during the preceding year was what mattered according to the bold statements of receipts and expenses, in cash and kind, that grew out of interrogations between the two parties. The spread of account-writing after around 1270 makes this more and more apparent, as the format became standardized and incorporated many more sub-divisions, which in turn facilitated the detailed itemization of individual products and projects.[82] This level of scrutiny can be traced into the late fourteenth century, at which stage it gradually disappears with the shift to demesne leasing, where the lessee was only responsible to his lord for payment of the basic farm.

The barn was an ideal setting to play out this tug-of-war between lord and peasant, since it stored the crops forming the manor's main source of livelihood. Although not usually the final storage point, the barn handled the harvest in its bulkiest form and obviously stood out more in the landscape than the much smaller granaries. As surviving sites show, barns are often the local point for the manorial complex, dominating the other buildings both with respect to size and relative positioning. In no case that I have examined is the barn inferior to any other standing work building. Surviving cow houses and stables, though large structures, are never comparable in stature. The remains of a magnificent byre at Shaftesbury Abbey's manor in Bradford-on-Avon, for example, showed that it was shorter and narrower than the barn, and rested at the bottom of the hill slope, along with all the other buildings, while the barn dominates the complex, standing at a higher level, proudly and confidently facing out on the town and the route that people used to approach the grange.

Harmondsworth provides further insight. A restless manor in the previous century, it had taken royal intervention in 1279 to bring the tenantry to heel after they had held out for four years in a dispute over customary services.[83] When Wykeham purchased the manor in 1392, he had a copy of the original customal that set out the services owed. John Laner, bailiff, ended his account in 1399 with a list of how much work he expected to exact from the tenants according to the customal, which amounted to 738 days.[84] Temporary sales of these works began in 1397-98 and increased throughout the fifteenth century, though not without some tension according to manorial records. Between 1420 and 1436, many petty disputes over timber and fishing rights

---

[81] Harvey, *Manorial Records*, p. 15.
[82] Ibid., pp. 16-7.
[83] *Victoria County History, Middlesex*, 2, ed. William Page, London, 1911, pp. 80-3; *VCH Middlesex*, 4, 12.
[84] Tim Mowl, 'The Great Barn at Harmondsworth', unpublished.

broke out, including armed raids on the college woods and heaths.[85] For its part, the rhythmic and accomplished design of the over-sized barn exudes an air of confidence and control. The timber framework is strong and simply built, and the smoothly curved, overbuilt arches which brace the side-aisles help to convey, better than any other barn, the aesthetic rhythm of the great Gothic cathedrals. The barn represented a prestige building project of the school that was aimed, however vainly, at subduing the restless tenantry of this manor.

That the barn did touch a general social consciousness is revealed in the closing two *passus* of William Langland's *Piers Plowman. Passus* XIX of the B-text introduces the reader to the final stage of the poem's climax: the stage is *Unitas*, and *Unitas* is a barn. Langland's allegorical description leaves the reader in no doubt as to how he perceived this barn. The frame, we are told, covers the whole as a roof:

> 'Before your grain', said Grace, 'begins to ripen
> Prepare yourself a house, Piers, to put your crops in.'
> 'By God, Grace', said Piers, 'you must give timber,
> And arrange for that house ere you go hence'.
> And Grace gave him the Cross, with the garland of
>   thorns,
> That Christ suffered on at Calvary for mankind's sake.
> And from His baptism and the blood that He bled on the
>   Cross
> He made a kind of mortar, and mercy was its name.
> And with it Grace began to make a good foundation,
> And wattled it and walled it with His pain and His
>   passion;
> And out of all Holy Writ he made a roof afterward;
> And he called that house Unity, Holy Church in
>   English.[86]

Langland was clearly familiar and in touch with his surroundings.[87] The timber barn he describes would not be out of place in the area of his possible home country, the Malvern Hills of Worcestershire, and it would be even more at

---

[85] *VCH Middlesex*, 4, 12.
[86] B.XIX.316-328, E. Talbot Donaldson, ed. and trans., *Piers Plowman*, New York and London, 1990, 224-5.
[87] Christopher Dyer, 'Piers Plowman and Plowmen: A Historical Perspective', *The Yearbook of Langland Studies*, 8 (1994), 155-7.

the B-text introduces the reader to the final stage of the poem's climax: the stage is *Unitas*, and *Unitas* is a barn. Langland's allegorical description leaves the reader in no doubt as to how he perceived this barn. The frame, we are told, covers the whole as a roof:

> 'Before your grain', said Grace, 'begins to ripen
> Prepare yourself a house, Piers, to put your crops in.'
> 'By God, Grace', said Piers, 'you must give timber,
> And arrange for that house ere you go hence'.
> And Grace gave him the Cross, with the garland of
>   thorns,
> That Christ suffered on at Calvary for mankind's sake.
> And from His baptism and the blood that He bled on the
>   Cross
> He made a kind of mortar, and mercy was its name.
> And with it Grace began to make a good foundation,
> And wattled it and walled it with His pain and His
>   passion;
> And out of all Holy Writ he made a roof afterward;
> And he called that house Unity, Holy Church in
>   English.[86]

Langland was clearly familiar and in touch with his surroundings.[87] The timber barn he describes would not be out of place in the area of his possible home country, the Malvern Hills of Worcestershire, and it would be even more at home in the London region, which Langland also knew well. What Langland does with Piers' barn at this stage for the climax of the poem is of particular interest. Once he has built his barn, Piers fills it with the harvest of his crop and then disappears away to plough, as a good plowman should. At this point (XIX.335), Pride seizes the opportunity of Piers' absence to attack Conscience and all Christians and the cardinal virtues. Conscience councils the brethren to flee to Piers' barn for safety, where Wit instructs Conscience to exhort the people to dig a deep moat around the barn, which they subsequently fill with penitent tears (XIX.354-380). Things start to go wrong, however, because Conscience asks the people that they render up their debts so that they might

---

[86]   B.XIX.316-328, E. Talbot Donaldson, ed. and trans., *Piers Plowman*, New York and London, 1990, 224-5.

[87]   Christopher Dyer, 'Piers Plowman and Plowmen: A Historical Perspective', *The Yearbook of Langland Studies*, 8 (1994), 155-7.

attain Piers' pardon and so be saved; *redde quod debes.*[88] The furor that this most Christian act incites suggests that Conscience has become the symbol of an ecclesiastical lord. The commoners are immediately suspicious: 'How's that? You counsel us to repay all we owe any man ere we go to Mass?' (XIX.391-392). But it is the brewer who is vehemently opposed and earns Conscience's wrath:

> 'Yes? bah!' said a brewer, 'I will not be ruled,
> By Jesus for all your jangling, by *Spiritus justitiae,*
> Nor by Conscience, by Christ, while I can sell
> Both dregs and draff, and draw from one hole
> Thick ale and thin ale; that's the kind of man I am!
> And I won't go hacking after holiness. Hold your
>   tongue, Conscience!
> Of *Spiritus justitiae* you speak a lot of nonsense'.[89]

We then hear from a vicar, who aspires to be the voice of the people (XIX.409-458). He is quite against paying up, since he feels he owes nothing. His argument ranges widely and a little disjointedly, but in essence he feels aggrieved and put upon. He is most annoyed by the sight of Church leaders, particularly cardinals, who enter the countryside demanding that the poor support them, even though the Church gives them nothing in return. Moreover, whenever Christians are killed or robbed, the Church is uninterested. This 'ignorant vicar' possibly voiced a popular grievance, since theirs was a poorly-paid lot which forever owed dues to the rector, be it a monastery, or, as in Swalcliffe Rectory, a college. Certainly, at the end of the *passus*, he leaves the sanctuary of the barn for his far-distant home. A lord then says his piece, and he too sees no reason to render up anything either (XIX.459-464). He is far more confident than the *lewed* vicar, and far less vulgar than the brewer. He turns Conscience's argument to his own advantage, saying that he happily takes everything that his auditor and steward offer him

---

[88]   'Come', said Conscience, 'you Christians, and dine,
         You who have laboured loyally all this Lenten time.
         Here is blessed bread, and God's body thereunder.
         Grace through God's word gave Piers power,
         Might to make it, so men might eat it after
         To help their health once every month,
         Or as often as they had need, those who had paid
         To Piers the Plowman's pardon *redde quod debes*'.
         B.XIX.383-390, Donaldson, *Piers Plowman*, p. 226.

[89] B.XIX.396-402, Ibid., p. 227.

as a matter of right. Finally, we hear from a king who browbeats Conscience into submission with the argument that the law (which the king heads) permits him to have anything he needs in return for being the earthly protector. Even more so than the lord, he owes nothing to anybody (XIX.465-479).

In *passus* XX, Pride's attack begins. Conscience calls his 'fools', or *foles*, together (presumably they were still outside the barn having just completed the moat), and urges them inside, while at the same time invoking the help of *Kynde* to do battle with the forces of evil (XX.74-79). It was, however, a forlorn hope because the *foles* remained adamant in their views. As the battle rages outside the barn, the Dreamer also becomes embroiled, and is cast among those who would not pay their debt. Old Age has rendered him bald, deaf, toothless, gouty, and finally so wizened that he is unable even to pay his marital debt (XX.183-211). Old Age calls him a 'lazy loafer', that most offensive of landless peasant (from an employer's perspective) who was able-bodied but unwilling to work.[90] The Dreamer cries out to *Kynde*, who advises him to find his way to *Unitas* and to learn to love. Throughout the poem, to love is to be faithful to one's lord, and it is readily apparent that it means the same here; the barn is not only a place of sanctuary, it is where one can show loyalty to one's lord. One must thresh and stack there, too, so it is not surprising in the end that *Unitas* fails to prevent the conquest by the forces of evil. Conscience, now portrayed as the constable or warden, still believes that the wounded will attain Piers's pardon if they do penance for their sins and render up what was owed (XX.304-308). However, some of the wounded did not favor this form of healing, and sent out for a doctor among the besiegers, 'that applied softer compresses' (XX.310). The upshot of this is the arrival of the frightful Friar Flatterer, whose compresses are far too soft, and the poem ends with Conscience giving up and leaving the barn in search of Piers.

The scenes around *Unitas* are comical, and the tenor of their delivery perhaps made the full questioning of the issues behind them more palatable to Langland's well-heeled audience. What is fascinating is his choice of a barn as the place to stage the climax of this epic work. As *Unitas* is Holy Church, a church—perhaps not a cathedral, but certainly a parish church—might appear to be a far more appropriate building to choose. His deliberate choice of a barn—Piers after all is a plowman—renders the passages as a readily appreciable critique of manorial authority. Nobody wishes to render what is owed. The producers of society, those who sweat and toil in the fields are almost silent, however. They would be among the mass of commoners who

---

[90] Dyer, 'Piers Plowman', p. 168.

quickly and sharply question Conscience's request, but we do not hear them making excuses; nor do we see them paying up. Langland concentrates on the controllers, the takers, such as the brewer, vicar, lord, and king, and pokes fun at them all; he rails against these people, not the honest (and somewhat idealized) peasant.[91] Dyer's belief that, 'Langland does not say much about the relationship between lords and peasants, nor of the antagonisms which, according to other sources, caused so much social contention',[92] appears unfounded because Langland does address this issue, but not as directly as we might expect. His approach is a discreet one that alludes to problems rather than addresses them head-on. The peasants in this case keep their mouths as tightly shut as possible, while their social and economic betters are shown to be brash and ultimately rather unattractive people.

Langland opted for the barn as a readily identifiable symbol of oppression in the landscape. His closing scene repeats one of the poem's central themes by contrasting a bygone utopia with the turbulent, impoverished world of the late fourteenth century which he depicts. For him, the spiritual idealism of Piers' barn has been destroyed and replaced by a structure that symbolizes the widespread corruption of the prevailing social institutions. The barn thus epitomizes social evil and is itself an impediment to progress. This episode gives a curious twist to the accepted view of technological development as a process of continual advancement by asking who specifically benefits from development and change.

CONCLUSION

There is still much to learn from a close study of medieval barn-building in England. The whole matter of structural design and its development will only be resolved by an in-depth study at national level. While my own work has so far focused on the lowlands in the fertile arable south, it is necessary to take account of barn-building traditions in more upland and northerly counties, where different agricultural systems based on livestock management predominated. Comparing designs across regions will better show how these structures performed their agricultural functions. This essay merely suggests different ways to look at barns that go beyond the traditional interest in their form and construction technique. For barns are cultural objects that reveal much about medieval rural society, as do other, more humble work

---

[91] These of course are to be distinguished from the wasters and loafers whom Piers upbraids in B.VI.115-198 when he tries to get them to plough his half-acre.
[92] Dyer, 'Piers Plowman', p. 166.

buildings.[93] Further work in this area will illuminate more fully features of daily life that are too often taken for granted.

TABLE 4.2

BUILDING EXPENSES AND ANNUAL REVENUES, SWALCLIFFE RECTORY, OXFORDSHIRE, 1397-99–1457-58.

YEAR[a]	BARN	OTHER BUILDINGS	RECEIPTS	EXPENSES	PROFIT
1397-99	£ 1 8d.	£ 7 11s. 6d.	£24		
1398-99			£53 6s. 11d.	£37 3s. 10d.	£53 2s. 9d.
1399-1400			£79 9s. 8d.	£52 8s. 4d.	£53 5s.
1400-01	£10 4s. 10d.		£80 7s. 11d.	£53 11s. 6d.	£52 8s. 5d.
1403-04	£ 6 8s. 9d.		£54	£54	£53[b]
1404-04	£20 6s. 8.5d.[c]	£11 2s. 4d.	£67 6s.	.	
1404-05	£ 5 19s. 4d.		£54	£54	£53 9s. 4d.
1405-05	£60 12s. 1d.	£2 13s. 10d.	£69 6s. 2d.	£58 18s. 1d.	
1405-06	£ 2		£52 3s. 4d.	£49 8s. 4d.	
1405-06	£11 18s. 9d.	£1 18s. 1d.	£13 4s. 4d.	£13 7s. 6d.	
1407-08	£ 5 18s. 4d.	19s. 3d.	£52 10s. 10d.	£50 3s. 4d.	
1408-09		6s. 1d.	£52 17s. 6d.	£40 16s. 2d.	£49 15s.
1409-10	£ 7 19s. 6d.	£26 10s. 3d.	£61 17s.	£61 4s.	£49 7s. 8d.
1409-10		£10 13s. 11d.	£51 3s.	£51 3s.	£48 18s. 6d.
1410-11		18s. 2d.	£50 13s. 4d.	£50 5s. 10d.	£41 14s. 2d.
1411-12		19s.	£50 17s. 6d.	£50 17s. 6d.	£49 9s. 3d.
1412-13	18s. 5d.	12s. 8d.	£50 10s.	£50 9s. 2d.	£48 12s. 3d.
1413-14			£50 10s. 10d	£50 10s.	£49 17s. 11d.
1414-15		4s. 2d.	£50 10s. 10d.	£49 14s. 10d.	
1415-16		1s. 1d.	£50 10s. 10d.	£50 3s. 4d.	
1416-17		£ 8 11s. 6d.	£51 10s.	£51 16s. 3d.	
1417-18		£ 1 8s. 3d.	£51 16s. 8d.	£51 15s. 9d.	
1418-19		4s.	£51 10s. 10d.	£51 10s. 10d.	
1419-20			£51 10s.	£51 10s.	
1420-21		3s. 9d.	£51 10s.	£51 10s.	
1421-22		1s. 11d.	£51 10s.	£51 13s. 6d.	
1422-23		15s. 7d.	£51 10s.	£51 10s.	
1423-24		10s.	£51 10s.	£51 10s.	
1425-26		7s.	£51 11s.	£51 11s.	
1426-27		10s.	£51 10s.	£51 10s.	
1428-29		2s. 6d.	£52	£34 4s. 7d.	
1429-30		17s. 8d.	£69 15s. 5d.	£60 11s. 8d.	
1430-31			£61 7s. 1d.	£46 4s. 2d.	
1431-32		£ 4 7s. 3d.	£67 6s. 3d.	£51 7s. 2.5d.	
1431-32		£ 7 17s. 2.5d.	£68 2s. 4.5d.	£44 2s. 0.5d.	
1432-33		5s. 9d.	£76 14s. 8d.	£64 17s. 7.5d.	
1433-34		£ 4 6s. 8d.	£64 10s. 11d.	£62 7d.	
1434-35		8s.	£52 16s. 8d.	£51 3s. 4d.	
1435-36		5d.	£54 6s. 8d.	£41 19s. 0.25d.	

Source: NCA 9150, 7001, 7002.

a. Accounting year typically began and ended on Michaelmas (29 September). See note 25.

b. This entire sum was set aside to meet the expenses of the barn-building.

c. This is a minimum figure since the account mentions a further £40 9s. 1d. as a shared expense with building works carried out at the neighboring presbytery of Shutford.

Christopher Dyer's recent paper, 'Sheepcotes: Evidence for Medieval Sheepfarming', *Medieval Archaeology* 39(1995) 136-64, has broadened the scope of inquiry.

Part II

Water, Wind, and Muscle

5

The Archaeology of Water Power in Britain
Before the Industrial Revolution

David Crossley

INTRODUCTION

The years 1965-85 saw an unprecedented number of excavations of water-powered sites in Britain. From these there was derived a body of evidence about the construction and installation of water wheels in the Middle Ages and the post-medieval centuries. A summary of the available evidence was published by Jernkontoret as a conference paper.[1] It is appropriate, ten years later, to ask how far the conclusions which were drawn have stood up to later assessment and to additional evidence, and what fresh areas of research have been explored.

Changes in the archaeological climate in Britain have reduced the number of opportunities for excavations of the kind that produced the evidence on which studies of medieval and later mills have come to rely. Rescue excavation, in advance of development, has become less frequent, partly due to economic recession, partly to the presumption, recently incorporated in the local authority planning process under national guidelines, that archaeological remains are a material consideration in deciding whether development should take place and, if allowed, how far damage to deposits can be mitigated. In addition, there has been something of a retreat from research excavation, a shift paralleled in other countries, arising from the debate over how far investigative destruction of archaeological deposits can be justified. Hence, for both these reasons the amount of new evidence from excavations is quite small, although what has emerged is of considerable value, and will be assessed here in the light of earlier results.

The changes outlined above have accompanied, perhaps caused, a shift towards non-destructive survey, and an interest in the interpretation of relict landscapes. This movement from site-specific studies has particular relevance

---

[1] D.W. Crossley, 'The Construction and Installation of Water Wheels: Medieval to Post-Medieval', *Medieval Iron in Society I*, Stockholm, 1985.

to the archaeology of water power. In the first place, it has encouraged the study of the complete mill complex, including the water-courses, weirs and dams, in addition to the details of mill buildings and water wheels. Further, it has led to more knowledge being gathered about the circumstances of construction of mills and, in particular, the extent to which the overcrowding of rivers led to compromises in design. In a wider context, there has been encouragement to study the mill as a component in early industrial landscapes. Over the Middle Ages and subsequent centuries, rural industry had an important economic role, and it is not hard to find examples of relationships between resources and manufacturing sites which stretched over considerable areas, as well as having linkages with settlements whose occupants relied on a combination of agrarian and craft sources of income. In Britain there is debate about the recognition and protection of landscapes of this kind. What can pose particular problems is the character and condition of landscape-components. It is common to find landscapes where individual sites or structures fall below the threshold of condition or technological significance which would justify statutory protection, but where the assemblage as a whole may be of major significance. The archaeology of water power is a key factor in this discussion.

THE WATER MILL IN THE LANDSCAPE

The need to place the water mill in its landscape context has led to surveys of entire valleys, in which available water supply is related to series of mills rather than to the individual site. Examples are the coverage of the 115 mill sites in the five river valleys of Sheffield,[2] one of which is shown in fig. 5.1. There has been comparable work on the streams of the Pennines in east Lancashire[3], and the current survey by Umpleby of the mill sites of the Dearne catchment, Yorkshire. There have also been surveys which have concentrated on mills with a particular function, such as the work of the Wealden Iron Research Group in south-east England, research on the water-powered tin-smelting sites of Devonshire by the Dartmoor Tin Research Group, and on lead-smelting sites in Derbyshire.[4]

The methods of storage and release of water varied according to the

---

[2] D.W. Crossley, J. Cass, N. Flavell and C.A. Turner, *Water Power on the Sheffield Rivers*, Sheffield, 1989.

[3] Greater Manchester Archaeology Unit, *Reversion Areas: the Identification and Survey of Relict Industrial Landscapes in Greater Manchester*, Manchester, 1990.

[4] H.F. Cleere and D.W. Crossley, *The Iron Industry of the Weald*, Leicester, 1985/new edition Cardiff, 1995; D.W. Crossley and D.T. Kiernan, 'The Lead Smelting Mills of Derbyshire', *Derbyshire Archaeological Journal* 112 (1992) 6-47; Dartmoor Tin Research Group: Newsletters, in progress.

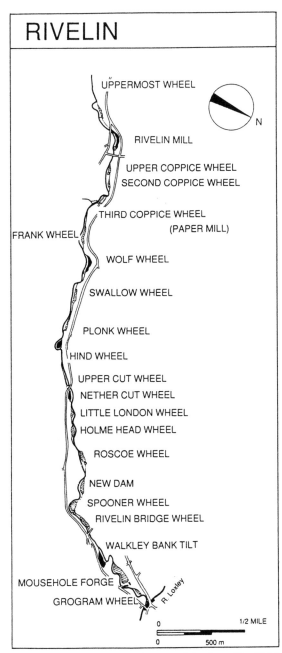

Figure 5.1. Mill-sites on the Rivelin, a tributary of the Don, Sheffield, Yorkshire.

quantities available. In districts of low rainfall or small catchments, such as the Weald of Kent, Surrey and Sussex, it was usual to construct a cross-valley dam to impound the entire flow for use by the mill. The surplus was released over weirs. The disadvantage of this form was the need to withstand the flow of winter streams, and surviving earthworks show that failure has occurred at the points where weirs or feeds to wheels were placed. In addition, the flow of water through the pond set up patterns of silting and scouring; regular dredging was required to maintain capacity. In the uplands, the common method of impounding water was to construct a by-pass system. The simplest variant was to place an undershot wheel in a short diversion from the stream, as a protection from storm water. This developed into a system whereby a channel or head-race provided not only a head of water sufficient for a breast, pitch-back or overshot wheel, but an element of storage capacity. Water would be fed into the race at a diversion immediately upstream from a weir, whose height regulated the level of water in the channel. The head-race, built along the contour of the valley side, was frequently enlarged to form a reservoir. Many weirs survive, often in positions used since the Middle Ages, and they are frequently the only means of locating mills shown on maps and surveys, particularly in urban areas where ponds and channels have been filled in and buildings demolished. Weirs, as Fitzherbert shows in his *Surveying* of 1523[5] were built of timber or stone, or both. The most durable were of pitched stones set between kerbs of stone or timber, sometimes strengthened by division into bays. The upper kerb of the surviving weir is generally of stone, often with iron fittings for washboards, planks set on edge to adjust the effective height of the weir. Observation of deteriorating weirs has called attention to the maintenance costs faced by millers: the force of flood water has removed stone kerbs, and damage quickly spreads across the structure. The risks could be reduced by building the weir at an angle to the flow: an excellent example can be seen at Cheddleton Mill, Staffordshire, where a pair of mills used for grinding flint for the pottery industry stand on the site of a medieval grain mill.

The methods of building earth dams have received insufficient archaeological attention in Britain, particularly as there is little documentary evidence. This contrasts with America, where Reynolds has demonstrated that a good deal of interest was being shown in traditional building methods in the first half of the nineteenth century when numerous small impounding schemes were being undertaken, of which diagrams survive.[6] In Britain, many dams dated from the Middle Ages, and in areas of later expansion such as the

[5] A. Fitzherbert, *Surveying*, 1523 (1767 edn), 91-5.
[6] T.S. Reynolds, *Stronger than a Hundred Men*, Baltimore, 1983, 128.

Weald, the Pennines or the Cotswolds, construction had largely ceased before illustrated publications on civil engineering topics became common. There have been few excavations where it has been possible to cut sections through dams. A limited examination was possible during excavations at Panningridge furnace, Sussex, where it was found that logs had been laid on the surface of a marshy valley before the earth dam was built.[7] At Maynards Gate, Sussex, a full section through the dam was possible, and this showed that the topsoil had been stripped, and that a bank of clay and sand had been laid without a foundation.[8] Field observation at other Wealden sites has supplemented the record. During pipe-laying at Sheffield Park it was noted that the iron-furnace dam had a stone core, probably dating from the sixteenth century. It was common in this area to raise dams, to maintain storage capacity in the face of continual silting. To contain the extra material, stone or timber revetments were built: stains of timber stakes were recorded at Maynards Gate, supporting blast-furnace slag, and at Westfield Forge there was a similar use of timber, on both sides of the dam. In the Pennines, ponds in by-pass systems have been shown from recent field observation to have been impounded by dams of some sophistication in design. This was due to their siting on hillsides. The method of construction can be seen from eroded bank-sections, which show the means used to minimise seepage back to the adjacent stream. A strong wall was built on the stream side, forming the core or facing of a bank which was built from earth or rock quarried from the hill side of the pond area. Puddled clay formed a waterproof bottom: it was vital that this should be kept intact, hence the need for preventing the growth of vegetation whose roots would damage the clay. This explains the frequent occurrence of sluices and shuttles which allowed the pond to be emptied for the removal of silt and weed.

An aspect of the topography of water power that may be overlooked is the means of returning water to the river. The design and maintenance of the tail race was of importance, preventing the rotation of the water wheel being impeded by what was known as 'back-water'. For the archaeologist, the tail race is important, often surviving as a drainage channel when little else can be seen. To prevent silting, not only was a good outflow needed, but the banks of

---

[7] D.W. Crossley, 'A 16th-Century Wealden Blast Furnace: Excavations at Panningridge, Sussex, 1964-70', *Post-Medieval Archaeology* 6 (1972) 42-68.

[8] O.R. Bedwin, 'The Excavation of a Late 16th/Early 17th-Century Gun-Casting Furnace at Maynards Gate, Crowborough, East Sussex', *Sussex Archaeological Collections* 116 (1977-8) 163-78.

the race had to be stable. At Caldecotte mill, Buckinghamshire,[9] excavations showed how vulnerable to erosion the banks had been, and at Chingley Forge, Kent[10] the tail race had been strengthened with large lumps of cinder from the forge hearths. In valleys where mills were closely sited, it was on occasion the practice for the tail race to feed directly into the pond belonging to the next mill downstream. Examples of this can be seen on streams in the Sheffield area, notably on the Rivelin, where numerous sites were developed for small cutlery-grinding wheels during the eighteenth century, giving a frequency so great that operation must have posed considerable problems. In fact, tandem mill siting could only be satisfactory where contiguous sites were in the ownership of one ground landlord, who could incorporate conditions of co-operative working into leases. In contrast, some mills had tail races of remarkable length. Sheffield has a number of examples which can be explained by constraints of property boundaries. Where it was difficult to accommodate dam systems that would give adequate head of water, it was sometimes feasible to dig a lengthy tail race. Hence it was possible to mount a wheel at a low level, and to design a long outlet channel with a fall that was adequate, but less steep than that of the river. The outfall was sited where the levels coincided. This can be seen at mills on the Rivelin, notably Third Coppice Wheel, Holme Head Wheel and Rivelin Bridge Wheel. In the case of the last two, the races are separated from the stream by walls of edge-set stone flags, while at Third Coppice, and at Rowell Bridge in the adjacent Loxley valley, there are earth and stone banks. There is a remarkable demonstration of such a system at another Sheffield mill, the Sharrow snuff-grinding mill on the Porter stream; the tail race leaves the wheel pit at such a depth that it was possible for it to be culverted below the river, which it joins at a distant point on the opposite bank, downstream.

WATER WHEELS AND THEIR INSTALLATION
Horizontal Mills

At the outset, it is helpful to survey what is known about mills at the end of the first millennium C.E., and to draw attention to the problems of this period. The Saxon water mill is epitomised by the example excavated at Tamworth, Staffordshire, and recently fully published.[11] Two phases, dating from the earlier and later years of the ninth century respectively, contained evidence for horizontal water-wheels. An earlier Saxon mill, at Old Windsor, Berkshire,

---

[9] M. Petchey and B. Giggins, 'The Excavation of a Late-17th-Century Water Mill at Caldecotte, Bow Brickhill, Bucks', *Post-Medieval Archaeology* 17 (1983) 65-94.
[10] D.W. Crossley, *The Bewl Valley Ironworks*, London, 1975.
[11] P.A. Rahtz and R.A. Meeson, *An Anglo-Saxon Watermill at Tamworth*, London, 1992.

has not been fully published; excavation showed that there had been a triple-wheeled vertical mill now thought from dendrochronological evidence to have been built late in the seventh century, and a later horizontal mill whose dating has not been satisfactorily resolved.[12] Evidence for horizontal mills in England is limited to these excavated examples and to the wheel-paddle found at Nailsworth, Gloucestershire, recovered without archaeological supervision and suggested as being earlier than 1327.[13] Otherwise, examples are known from Ireland, summarised by Rynne,[14] and from the Scottish islands, notably Orkney and Shetland, where standing structures survive.[15] There is ample continental evidence, notably the range of Mediterranean examples summarised by Rahtz,[16] in a tradition which in the nineteenth century was to evolve into the early water turbine. What is not clear is how long the horizontal mill continued to be constructed or used in England. Apart perhaps from the Nailsworth example, whose dating is in doubt, there is no indication of this type of mill after the Norman conquest. This remains an important issue, deserving further research, for it is an open question as to how many of the 5,624 mills indicated by the Domesday survey were of this type.

Vertical Mills
The evidence for the construction of vertical-wheeled mills comes from a series of excavations of sites in use from the twelfth to the eighteenth centuries. These have shown that in Britain there was a tradition of construction which lasted over the entire period, and that any division between medieval and post-medieval is artificial. The major excavations of mills which were in use before 1500 comprise work on the grain-mill at Batsford, Sussex,[17] working in the fourteenth century, the forge-mill at the Cistercian abbey of Bordesley, Worcestershire, whose phases cover the period between the end of the twelfth century and the end of the fourteenth,[18] and the fourteenth-century mill at Chingley, Kent, where there is debate over the evidence for use as a

---

[12] B. Hope-Taylor, 'Note', *Medieval Archaeology* 2 (1958) 183-5; Rahtz and Meeson, *An Anglo-Saxon Watermill*, p. 156.

[13] Ibid, 102.

[14] C. Rynne, 'The Introduction of the Vertical Watermill into Ireland: Some Recent Archaeological Evidence', *Medieval Archaeology* 33 (1989) 21-31.

[15] G.D. Hay and G.P. Stell, *Monuments of Industry,* London, 1986, 8-10.

[16] P.A. Rahtz, 'Medieval Milling', in D.W. Crossley, ed., *Medieval Industry*, London, 1981, 1-15.

[17] O.R. Bedwin, 'The Excavation of Batsford Mill, Warbleton, East Sussex', *Medieval Archaeology* 24 (1980) 187-201.

[18] G.G. Astill, *A Medieval Industrial Complex and its Landscape: the Metalworking Watermills and Workshops of Bordesley Abbey*, London, 1993.

Figure 5.2. Excavated fragment of a water wheel in the second of two sixteenth-century wheel pits at Panningridge furnace, Sussex (scale 1 m).

Figure 5.3. Chingley forge, Kent: the sixteenth- to seventeenth-century wheel pit whose timbers suggest that two wheels were accommodated, side by side. The timbers of the fourteenth-century wheel-pit sides can be seen below and to the left of the later structure. The super-imposed horizontal timber crossing the foreground belongs to a later wheel pit, illustrated in fig. 4 (scale 2 m).

grain mill and as a hammer-forge. The mills at Batsford and Chingley were approximately twenty miles apart, and showed remarkable similarity in the details of their carpentry. For the post-medieval period, evidence comes from excavations of industrial mills, several being connected with iron smelting. In the Weald of south-east England, water-wheel fragments have been excavated at the site of Batsford blast furnace, Sussex,[19] Chingley furnace, Kent,[20] Maynards Gate furnace, Kent,[21] Panningridge furnace, Sussex,[22] Scarlets furnace, Kent,[23] and, in Yorkshire, at Rockley bloomery[24] and blast furnace.[25]

No post-medieval wheel fragments have been excavated at grain-mill sites, although the excavation at Caldecotte, Buckinghamshire, has provided important information about wheel-installation. The tradition of wheel-construction in wood was maintained until the end of the eighteenth century, when the use of iron parts for water wheels and for control mechanisms and penstocks brought a new precision to design and construction. It was at this time that competition from the steam engine led to scrutiny of the viability of water power, evidenced by the numerous calculations by millwrights, engineers, and surveyors of the power available at existing mill sites. Not only were questions asked about alternative sources of power, but also about the most advantageous use of ground occupied by mills, their water courses and reservoirs. In urban areas, rising site-values could lead to abandonment of mills in favour of more profitable uses of land.

The Installation of the Vertical Wheel
In the excavated examples, the water-wheel was set in a pit or casing, of timber or stone, which became increasingly sophisticated over the period. The timber examples have frequently survived, in damp valley-bottom conditions, although there is a sharp cut-off in the survival of wood at the lowest point in the long-term range of movement of the water-table. This is well illustrated in fig. 5.2, which shows the later of the two wheel pits and wheel fragments

---

[19] O.R. Bedwin, 'The Excavation of a Late-16th-Century Blast Furnace at Batsford, Herstmonceux, Sussex', *Post-Medieval Archaeology* 14 (1980) 89-112.
[20] Crossley, *The Bewl Valley Ironworks*.
[21] Bedwin, 'The Excavation of a Late-16th/Early 17th -Century Gun-Casting Furnace'.
[22] Crossley, 'A 16th-Century Wealden Blast Furnace'.
[23] D.W. Crossley, 'A Gun-Casting Furnace at Scarlets, Cowden, Kent', *Post-Medieval Archaeology* 13 (1979) 235-49.
[24] D.W. Crossley and D. Ashurst, 'Excavations at Rockley Smithies, a Water-Powered Bloomery of the 16th and 17th Centuries', *Post-Medieval Archaeology* 2 (1968) 10-54.
[25] D.W. Crossley, 'The Blast Furnace at Rockley, South Yorkshire', *Archaeological Journal* 152 (1995), 381-421.

excavated at Panningridge furnace. the pit-sides could provide support for the wheel-bearings; several examples, of which the sixteenth to seventeenth-century phase at Chingley forge is the most striking (fig. 5.3), were constructed as integral parts of the structure of the mill building. Even so, a cautionary note is struck by the illustration in the fourteeenth-century Luttrell Psaltar, which shows a wheel bearing mounted on posts, an arrangement that would leave few traces in the archaeological record. In some of the examples excavated, the box frame not only housed the wheel; it also formed the first section of the tailrace. The base timbers of the frame were set in a ditch excavated at right angles to the dam. Substantial posts were tenoned to this structure, and to these were secured the planks that lined the wheel pit. On the ends of the vertical members there was mounted a second frame, which formed the top of the wheel pit, supported the end of the pentrough that fed water from the pond to the wheel, and in several cases was integrated with the timber sill beams that made up the ground frame of the mill. Such construction was used at Batsford and at Chingley. The installation at Bordesley Abbey was less complex, the wheel-pit and tail-race timbers forming a unit separate from the foundations of the mill.

Box-frame wheel structures in a similar tradition have been excavated at sixteenth-century ironworks in southern England. At Panningridge furnace, built in 1542, the wheel pit comprised a simple three-sided box structure, framed with square-section oak members, with horizontal boards nailed to the exterior. This structure was not integrated with timbers in the furnace bellows-house. The wheel pit at Chingley furnace, Kent (ca. 1560) was similar, although it incorporated the timbers of a culverted tailrace that ran beneath the casting floor of the furnace. It is possible that the Panningridge installation had originally been similar, but that the tailrace extension had been destroyed during rebuilding in the second half of the sixteenth century. The Chingley wheel pit differed from earlier examples in having a planked floor, which would assist the flow of water. In the earlier wheel pits there was no trace of planking, and exposed cross-sleepers in the pit bottoms would create turbulence. A notable variant within this group of box-frame wheel pits was the late-sixteenth-century double wheel pit at Chingley forge, where it appears that two wheels were mounted on a common shaft, the wheels turning in adjacent casings (fig. 5.3).

Some mills of the period possessed stone rather than timber wheel pits. A sixteenth-century example is Maynards Gate furnace, and the stone race at Chingley forge dated to the seventeenth century. The reasons for the choice of materials must lie in the willingness of the owner or tenant of a mill to invest for the long term, and it is conjectured that there must be some consideration such as length of lease or likelihood of profitable use that lies behind the decision to build in stone. The quality range of such structures is considerable, from the relatively crudely-built walls of the pits at Maynards Gate and Chingley to the high standard of ashlar masonry at a further Wealden furnace site, the mid-seventeenth-century blast-furnace at Scarlets, Kent.

A problem that remains to be solved is at what period millwrights began to construct the characteristic breast-wheel pit. There were many situations where the head of water was insufficient to operate an overshot wheel, and where the efficiency of a wheel fed at a lower level could be assisted by the use of a pit with a curved rear profile, concentric with the wheel and designed to hold the water between the paddles or buckets for the maximum proportion of rotation. It has been accepted that this feature became common in the late eighteenth and the nineteenth centuries, and that such wheel pits were generally built in stone. However, the timber wheel pit associated with the final phase of operation at Chingley forge, ending ca. 1750, used a timber breast (fig. 5.4), and this raises the question of how far back this design can be taken. What may be a crude attempt at such a form, in stone, was excavated at Maynards Gate, in a late sixteenth-century context. The back wall of the pit had a slightly concave slope, which could not have been a precise fit with the wheel.

ARCHAEOLOGICAL EVIDENCE FOR WATERWHEELS
The medieval and early-post-medieval mills which have been excavated over the last thirty years have provided an important series of vertical-wheel fragments. Those illustrated in fig. 5.5 all belong to a tradition of construction in oak, and the majority appear to have been overshot. Almost all come from south-east England; the exception, from Yorkshire, shares features with the southern group, suggesting a widely-spread tradition of design. The wheel-fragments excavated comprise spokes, sole-boards, side-boards and buckets. There are no surviving hub-fragments, and only at Chingley furnace was there a piece of a wheel-shaft. This comprised the length of the shaft that operated the furnace-bellows, indicated by mortices for cams. Details of how wheel-spokes were fitted to the shaft were not clear, but it was evident from other components that English wheels of this period were of compass rather than clasp design, contrasting with those from the continent of Europe illustrated by Agricola.[26] Continental wheels were of widely differing design, as can be

<hr />

[26] G. Agricola, *De Re Metallica*, Basle, 1556; eds., H.C. and L.H. Hoover, New York, 1950, 183-206.

Figure 5.4: The final wheel pit at Chingley forge, out of use by the middle of the eighteenth century (scale 3').

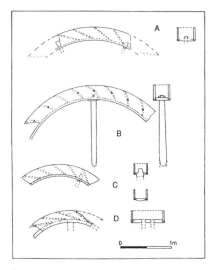

Figure 5.5. Diagrams of excavated water-wheel fragments; all are overshot. A) Rockley forge, Yorkshire, early seventeenth century. B) Chingley furnace, Kent, sixteenth century. C) Batsford furnace, Sussex, sixteenth century (after Bedwin 1980). D) Chingley forge, Kent, seventeenth to eighteenth century.

seen not only in Agricola's drawings and those those illustrating the works of Biringuccio and in Flemish landscape paintings. Reynold has discussed the reasons for diversity, emphasising the need for a supply of timber of substantial cross-sections for wheels of compass patterns, whereas the slender spokes and sole-board planks shown in some continental sources could be derived from wood of younger growth.[27]

It is possible to re-construct the details of design from excavated fragments. The wheel shaft, as Fitzherbert wrote in 1523, was a costly and substantial timber, normally oak, as was the case at Chingley furnace, where the only piece of a wheel shaft so far excavated was found. Fitzherbert describes the bearings on which wheel shafts were supported: he refers to the bearing-journals as gudgeons, running in bearings of bell-metal or of stone. Gudgeons are illustrated in eighteenth-century sources, and their design is discussed by Reynolds.[28] Iron pins were used, driven into the shaft-ends, the latter being bound with iron hoops to prevent splitting. The sixteenth-century Sussex ironworks accounts of the Sidneys contain references to replacement both of these hoops and of gudgeons. Lubrication was with tallow, and some millers directed water onto the bearings to keep them cool. Bearings were mounted in blocks of wood or stone: an example of the former was found *in situ* at Chingley furnace, and a stone bearing-block was found in the eighteenth-century wheel pit at Rockley furnace.

The best examples of wheel-spokes were recorded at Batsford Mill, Panningridge Furnace and Chingley Furnace. The numbers of spokes can often be estimated: the most certain case was at Batsford, where the survival of two adjacent spokes showed that there must have been a total of four. At Panningridge and Chingley the length of sole-boards adjacent to spokes showed that there could have been no more than six spokes in either case; at Rockley forge, Yorkshire, mortices in the sole-boards showed that both the wheels had eight spokes.

The wheel spokes were tenoned to the sole boards, British examples of the latter being curved members, shaped by adze, and up to 2 m in length. So far, there have been no examples of the cross-planked soles illustrated by Agricola. The shrouds, or side-boards, termed *compost* (compass?) boards by Fitzherbert, were nailed to the sides of the soles. The greatest variation in design amongst excavated wheel fragments has been in the design of paddles and buckets (fig. 5.5). These are the key to understanding the configuration of

[27] T.S. Reynolds, 'Clasp Arms Versus Compass Arms in Water Wheels: Regional Patterns or Timber Problems?', in *Medieval Iron in Society II*, Stockholm, 1985, 61-5.
[28] Reynolds, *Stronger than a Hundred Men*, pp. 160-61.

Figure 5.6. Shepherd Wheel, Sheffield, Yorkshire: A blade-grinding mill dating from the sixteenth century, subject to many subsequent alterations. The water wheel is largely of iron, but the pentrough is of timber. The control system is visible at the left-hand end of the pentrough.

wheels and their water supply, for there are marked differences between the paddles of undershot wheels and the buckets of overshot wheels. To distinguish between the latter and the equivalent parts of pitch-back wheels is less simple, for their buckets are essentially similar, the wheels rotating in opposite directions. Identification depends on being certain that the wheel as excavated is in the position in which it operated. It will be seen from the diagrams that there are detailed differences in the form and fixing of wheel-buckets. In most cases the bucket boards are straight, but the sixteenth-century Chingley furnace wheel is important for its curved boards, precursors of those used in the eighteenth century when a more theoretical approach to design becomes apparent. This wheel is also striking for its use of dowels rather than nails, and the precision with which the dowel-holes had been drilled through the width of the boards.

There is little excavated evidence for methods of controlling the speed of water wheels. Observation of mills surviving into the present century shows that for overshot and pitchback wheels the most common system is a rising shuttle (penstock) at the end of a pentrough, the timber or cast-iron box which was in effect an extension of the pond. The penstock is lifted and lowered by a rack-and-pinion system operated by a series of levers from within the mill (fig. 5.6). This arrangement is dependent on the use of iron parts, and has the advantage of control from the work-place. In the eighteenth century and before, we may infer that penstocks were lifted and lowered without any intermediate mechanism, and were fixed in place by pegs through a vertical shaft attached to the penstock boards. Early post-medieval illustrations suggest that the water-flow was not necessarily regulated close to the wheel, and that the shuttle could be sited at the pond end of a trough which directed water onto the wheel. The Sidney accounts for Panningridge furnace use the term 'shoot' for this item, and a re-used example was found built into the wheel-pit at the early-eighteenth-century furnace at Pippingford, Sussex.[29] Apart from this evidence, nothing is known of pentrough development, although at Chingley, both at the forge and furnace sites, there were traces of posts rising from the backs of wheel pits, which were likely to have supported troughs of some kind. At the mid-sixteenth-century blast furnace at Chingley a beam-slot was excavated on the top of the dam, aligned in the direction of the wheel pit and presumed to have supported a trough or shoot. We know less about the regulation of supply to wheels fed from a lower level. There was no trace of any sluice or shuttle for the undershot wheels at the double corn mill of seventeenth-century date at Caldecotte, nor was it clear how the water

---

[29] D.W. Crossley, 'Cannon Manufacture at Pippingford,Sussex: The Excavation of Two Iron Furnaces of c. 1717', *Post-Medieval Archaeology* 9 (1975) 1-37.

culverted through the dam to the undershot or low breast wheel at Ardingly forge (later fulling mill) was regulated.[30] Nineteenth- and twentieth-century survivors of this form of mill, such as at Cheddleton, have lifting penstocks in the race that feeds the wheel, operated by rack and pinion mechanisms similar to those used in the late-type pentroughs for overshot wheels.

CONCLUSION

The excavations of the last thirty years have provided a basic corpus of information about the details of how water was harnessed to provide power for industry in the Middle Ages and beyond. We are able to visualise and confirm descriptions such as those of Fitzherbert, and to assess how far illustrations such as those from the *Luttrell Psaltar*, Biringuccio's *De Pirotechnia*,[31] Agricola's *De Re Metallica* and the landscape paintings of the Breughels and their contemporaries compare with practice in Britain. As explained at the start of this paper, it is unlikely that any great volume of excavated evidence, as was revealed in the period 1965-85, will again be forthcoming, but even if it were, it is unlikely that more than detailed additions to knowledge would be gained. The change to a less interventionist archaeological strategy is typical of much current research. In the 1990s a wide-ranging reappraisal of sites and landscapes of all periods is taking place, spurred by the need to ensure adequate protection for sites and monuments. This is being led by the Monuments Protection Programme (MPP), undertaken by English Heritage (the Historic Buildings and Monuments Commission for England), the statutory advisory body to government. Landscape evidence from MPP field surveys is doing far more than producing lists of priorities for protection: it is providing material for an overall re-assessment of archaeological landscapes, in which evidence for early industry is an important element. It is in this particular field of study that the evidence for the use of water as a source of power has an important part.

---

[30] O.R. Bedwin, 'The Excavation of Ardingly Fulling Mill and Forge, 1975-76', *Post-Medieval Archaeology* 10 (1976) 34-64.

[31] V. Biringuccio, *De Pirotechnia*, 1540, eds. C.S. Smith and M.T. Gnudi, Cambridge, MA, 1959.

6

'Advent and Conquests' of the Water Mill in Italy*

Paolo Squatriti

In the November 1935 issue of his soon-to-be famous journal, then called
Annales d'histoire économique et sociale, Marc Bloch published a subtle and
influential paper on the 'advent and conquests' of the water mill in medieval
Europe.[1] Although he was by no means the first to treat the history of mills in
medieval Europe, everyone since then who has approached this important
subject has cited Bloch's insightful work reverently, and indeed it still
provides an excellent point of departure for analysis of medieval milling
technologies and their social impact. The influence of Bloch's thought on this
subject, even today, is such that a summary of his main positions is worth
making. The great French medievalist traced the orderly stages of the
diffusion of the water-powered mill after its invention in the eastern
Mediterranean region in late Hellenistic times. For him, the spread of this
technology was best explained by conceiving a series of progressively wider
circles emanating as time passed, like the proverbial ripples of a stone cast
into still water, northwards and westwards from the inventive East. After
analyzing the ancient literary sources, Bloch proposed that hydraulic
technology in general, and hydraulic milling in particular, were little
developed in the ancient world which invented them. The slow, hesitant
spread over several centuries in Antiquity contrasted with the sudden,
innovative diffusion of water milling in medieval Europe, by which he
primarily meant France and England (in a footnote he excused himself for not
considering other areas on account of the inaccessibility of their medieval
documents).

Bloch further postulated that this medieval success story was related to
new conditions of labor prevailing by the ninth and tenth centuries. Whereas

---

* This study was much improved by the generous criticism of John Muendel and Michael
Allen, to whom I am grateful. They are, of course, not responsible for its remaining faults.
[1] Bloch, 'Avènement et conquêtes du moulin à eau', *Annales d'histoire économique et
sociale* 7 (1935) 538-63.

an abundance of slaves and widespread snobbishness had made Roman landowners uninterested in labor-saving devices like water mills, avid medieval lords perceived that such devices would waste less of the sweat of their serfs' brows and increase the opportunities for extracting surplus from them. For Bloch, in other words, acceptance of the technology of water mills was essentially a product of the demographic and economic conditions on the great manorial estates which lords accumulated especially in the Carolingian epoch. In the same issue of the *Annales* he reiterated his interpretation in another essay on 'medieval inventions', showing how a celebrated book by Lefèbvre de Noëttes, published in 1924, had put the cart before the horse. Lefèbvre de Noëttes had claimed that medieval technology liberated manpower and caused the end of ancient slavery, but Bloch demonstrated that medieval water mills were on the contrary a result of the lack of available labor.[2]

Largely thanks to Bloch's lucid analysis, water-powered mills have continued to attract much attention from historians since 1935; in fact, few water-related topics attract more attention from students of the Middle Ages.[3] Scholars of the ancient world have likewise examined the issue of hydraulic technologies raised by Bloch's work.[4] Some modern historians of Antiquity, worried by the Romans' apparent regrettable indifference to technological progress, have worked to correct the view that water milling was a freakish

---

[2] 'Les "inventions" médiévales', *Annales d'histoire économique et sociale* 7 (1935) 634-43. Bloch's 'Technique et évolution sociale: réflexions d'un historien', *Europe* (1938) 23-32 repeated his ideas.
[3] Some of the most engaging recent studies include C. Dussaix, 'Les moulins à Reggio d'Emilie au XIIe et XIIIe siècles', *Mélanges de l'École Française de Rome* 91 (1979) 113-47, who goes beyond technological matters; C. Parain, 'Rapports de production et dévelopement des forces productives: l'example du moulin à eau', *La pensée* 119 (1965) 58-66; M. Del Treppo, *Amalfi medioevale*, Naples, 1977, 45-9. Technology is the focus for J. Muendel, 'The Horizontal Mills of Medieval Pistoia', *Technology and Culture* 15 (1974) 206-8; L. Hunter, 'The Living Past in the Appalachians of Europe', *Technology and Culture* 8 (1966) 547-63; B. Gille, 'Le moulin à eau', *Techniques et civilisations* 3 (1954) 1-4; G. Beggio, 'Navigazione, trasporto, mulini sul fiume: i tratti di una tipologia', in *Una città e il suo fiume*, ed. G. Borelli, Verona, 1977, 550-51. Among general studies the standard R. Forbes, 'Power', in *A History of Technology*, v. 2, ed. C. Singer, Oxford, 1957, 590-610, is not superceded by B. Gille, *History of Techniques*, v. 1, New York, 1986, 341-3, 451-9.
[4] The best of the classical reference works is *Der kleine Pauly*, v. 3, Stuttgart, 1965, 1446-7. Classical historians' studies include M-C. Amouretti, 'La diffusion du moulin à eau', in *L'eau et les hommes en Méditerranée*, ed. A. De Reparaz, Paris, 1987, 16-20; D. Simms, 'Water-driven Saws, Ausonius, and the Authenticity of the *Mosella*', *Technology and Culture* 24 (1983) 643; Ö. Wikander, 'Water Mills and Aqueducts', in *Future Currents in Aqueduct Studies*, ed. A. Hodge, Leeds, 1991, 131-36; K. White, *Greek and Roman Technology*, Ithaca, 1984.

oddity in the Roman Empire before the Germanic migrations.[5] They have brought much epigraphical, archaeological and other evidence to light since 1935, to prove that water pushed many more Roman millstones than Bloch allowed. Their evidence has included tombstones celebrating the lives of 'water mill engineers', carved in Anatolia, and the Phrygian *vicus* of Orkistos which congratulated itself in an inscription for having many water mills, a sign of civility. Considering that mills were often fragile structures built alongside the temperamental streams of the Mediterranean and hence liable to destruction by flooding, and considering that relatively few classical archaeologists look beyond the urban centers and *fora*, to the rural areas where most mills operated, the amount of archaeological data on ancient water mills recovered in the past half century is most impressive. Water mills at Arles and Tournus in Gaul, Venafro near Naples, in the baths of Caracalla at Rome, and at several sites in the eastern Mediterranean, have revealed their shapes to excavators.[6]

This fresh archaeological material requires a reconsideration of Bloch's contention that the silence of Roman writers meant that the water mill was little known and unappreciated in the ancient world. This divergence between the reticent written sources upon which the creator of the *Annales* relied and the noisy churning of hundreds of water wheels on ancient Mediterranean streams attested to by archaeology is instructive. With the exceptions of an anonymous Augustan epigrammist and of Ausonius, whose *Mosella* (verses 362-4) is anomalous in more ways than one, the Roman men of letters who wrote the texts Bloch combed through with such acumen did not think water mills fitting subjects for their compositions. The likelihood of a Roman or any other poet waxing eloquent on the grinding of grain is, of course, small, and those cursory discussions of water power that do occur in Roman texts are in works such as the technical manuals of Vitruvius or

---

[5] See R. Holt, *The Mills of Medieval England*, Oxford, 1988, 3-5. White, *Greek and Roman*, 56, 196-201, attempted to show the ancients' openness to milling technologies; also Wikander, 'Water Mills', pp. 143-4; D. Lohrmann, 'Travail manuel et machines hydrauliques', in *Le travail au moyen âge*, ed. J. Hamesse, C. Muraille-Samaran, Louvain, 1990, 36-42; G. Comet, *Le paysan et son outil*, Rome 1992, 390-96; Amouretti, 'La diffusion', pp. 19-23; J. Oleson, *Greek and Roman Water-Lifting Devices*, Toronto, 1984, 375-6; M. Nordon, *L'eau conquise*, Paris, 1991, 103-4.

[6] Tombstones: Forbes, 'Power', p. 600. Mills' evanescence: Amouretti, 'La diffusion', p. 14. For a revealing notice of 'medieval mills' near Rome that baffled classical archaeologists, see S. Judson, A. Kahane, 'The *ager veientanus*', *Papers of the British School at Rome* 36 (1968) 177-8. Forbes, 'Power', pp. 596-9 and *Studies*, p. 93 mentions archaeological examples. Caracalla: H. Manderscheid, 'La gestione idrica delle terme di Caracalla: alcune osservazioni', *Les thermes romains*, Rome, 1991, 50-52. Wikander, 'Water Mills', pp. 14-5, 28 summarizes the issues well.

Palladius, and in the elucubrations of the encyclopaedist Pliny.[7] In addition to the paucity of literary allusions to mills, the fact that classical Latin did not distinguish semantically between types of mill renders the study of even these few texts arduous, and it is often difficult to determine what type of mill—whether horizontal- or vertical-wheeled, for example—Roman writers refer to when they do refer to one at all.[8]

Bloch's underestimation of the degree to which Roman societies exploited water to mill grain is thus related to his forced reliance on a literature that was prone to overlook such mundane technicalities even as it built agrarian myths of virtuous and tidy countrysides. The unwillingness of ancient writers to treat hydraulic technology probably did stem, as Bloch perceived, from a cultural predisposition enshrined in these literary conventions. Despite the *Georgics* and Cincinnatus who returned to his plough, Roman aristocrats, whose culture Roman authors reflected, came to dislike physical labor and anything that smacked of it, as most technologies did.[9] Ever since Plato had condemned hand-work as inferior to mind-work, classical thinkers considered *téchne* only when it was a purely abstract intellectual exercise divorced from lowly pragmatic applications.[10] The palpable distaste of Roman litterati for manual work thus conditioned the writings upon which Bloch drew in plotting the 'advent and conquests' of the water mill.

Nevertheless, outside of Rome's learned books, the technology of milling with water spread nicely. Evident in the archaeological record, their

---

[7] Pliny (*Nat. Hist.* 18.95), Palladius (*Op. Agric.* 1.41), and Vitruvius (*De Arch.* 10.5.2) discuss water mills. See Gille, 'Le moulin à eau', p. 1; Oleson, *Greek and Roman*, p. 375. On the fourth-century developments, Amouretti, 'La diffusion', pp. 19-20; A. Guillerme, *Le temps de l'eau*, Seyssel, 1983, 93; Parain, 'Rapports de production', p. 62; Forbes, 'Power', pp. 590, 600; Wikander, 'Water Mills', p. 28.

[8] Several terms were used indiscriminately, as C. Koehne, *Das Recht der Mühlen bis zum Ende der Karolingerzeit*, Breslau, 1904, 12-3 explains. Archaeologically, undershot and overshot vertical-wheeled mills are attested; horizontal mills are not, though they were probably more common. On early medieval molinological terms, P. Aebischer, 'Les dénominations du 'moulin' dans les chartes du moyen âge italien', *Bulletin DuCange* 7 (1932), 49-109. Simms, 'Water-driven Saws', pp. 635-43 exemplifies the difficulty of making technological sense of poetry. See also Forbes, 'Power', p. 597.

[9] T. Reynolds, *Stronger Than a Hundred Men*, Baltimore, 1983, 32; Nordon, *L'eau conquise*, p. 109. An example of authors who link Rome's 'prehistoric' level of technology to this attitude is F. and J. Gies, *Cathedral, Forge, and Waterwheel*, New York, 1994, 17-35.

[10] There is the interesting exception of military technology, but this point is well made by L. Cracco Ruggini, 'Progresso tecnico e manodopera in età imperiale romana', in *Tecnologia economia e società nel mondo romano*, Como, 1980, 47-51, 59-66. Plautus, who was said to have earned a living grinding grain while his compositions were unsuccessful, was an incongruity: Eusebius-Jerome, *Chronicon*, ed. R. Helm, Berlin, 1984, 135-6 (h).

wide distribution is also confirmed in several ancient and late antique documents, which eluded Bloch despite his thorough trawl through the literature.[11] The *Talmud*, for instance, discussed the legal implications of grinding grain with water on the Sabbath in Roman Palestine, while in Diocletian's *Edict of Prices*, where taxes on water mills were assessed, the legislator considered these machines altogether normal.[12] The familiarity of Roman society with water mills already in the third century, implied by Diocletian's laws, is corroborated by other evidence. The *Theodosian Code*, revealing the nervous attention Roman rulers dedicated to the supply of grain for the *Urbs*, frequently referred to mills and milling. For the fourth-century legislators mills were water-driven machines, and the emperors specifically favored hydraulic mills in their provisions.[13] These prescriptive texts presumed the presence and full acceptance of water mills in Roman society, and mirrored a world heavily dependent on flowing water at least for transforming grain into flour.

In the Italian peninsula, water mills were thus a fixture by C.E. 500. Cassiodorus, reflecting a common preoccupation of sixth-century monastic legislators, had urged monks to consider the suitability of local streams for milling before they elected a site for their monastic life, while his contemporary Procopius had recorded the stratagem of Belisarius, besieged in Rome, who deployed boat-mills on the Tiber in an effort to secure the city's supply of flour when the Ostrogoths cut water supplies to the aqueduct-fed mills on the Janiculum hill.[14] As these examples suggest, both in the religious

---

[11] Late Roman water mills: Reynolds, *Stronger*, p. 30; Comet, *Le paysan*, p. 405; P. Dockès, 'Formes et diffusion d'une innovation technique; le cas du moulin hydraulique', in *Forme ed evoluzione del lavoro in Europa, XIII-XVIII secc.*, ed. A. Guarducci, Florence, 1991, 125, 128.

[12] *Talmud: Shabbath* 1, tr. I. Epstein, London, 1938, 74 (Forbes, 'Power', p. 614, suggests Roman rule brought such mills to Palestine). *Diokletians Preisedikt* 15.52-5, ed. S. Lauffer, Berlin, 1971, 147.

[13] *Codex Theodosianus* 14.15.4, (C.E. 398), ed. T. Mommsen, 1.2 Berlin, 1905, 790-1. See also Cracco Ruggini, 'Progresso tecnico', pp. 53-4.

[14] Cassiodorus, *Institutiones* 1.29, (like *Regula Benedicti* 66), ed. R. Mynors, Oxford, 1961, p. 73 did not need to explain what he was talking about to his audience. Procopius, *Guerra Gotica* 1.19, ed. D. Comparetti, Rome, 1895, pp. 141-5 (see also Cassiodorus, *Variarum Libri XII* 3.31, ed. T. Mommsen, *MGH Auctores Antiquissimi*, 12 [Berlin, 1894], p. 95). The Janiculum water mills (evidently overshot, vertical-wheeled mills) had been central for Rome's *annona* since at least 389 when, according to the ecclesiastical historian Socrates (*Historia Ecclesiastica* 5.18), Theodosius closed the disreputable 'sweat' mills: Cracco Ruggini, 'Progresso tecnico', pp. 53-4 further suggests that their appearance in Prudentius' *Contra Symmachum* 2.950, ed. M. Lavarenne, Paris, 1963, 190, is the earliest reference to them, but these are not explicitly water-powered mills.

and in the secular worlds of the sixth century, flowing water still attracted attention whenever the problem of grinding grain arose. The water mill was popular on large estates but also in other, urban contexts, and machines with very different technical requirements, with horizontal and vertical paddle wheels, coexisted. Although the declining Roman state no longer maintained sophisticated milling technologies like those of the geared vertical-wheeled type, other water mills, better suited to the new, more rural economies of Italy remained popular. Thus in 643, when the Lombard king Rothari drew on custom to address problems created by the successes of mills in the previous centuries, he could expect everyone to understand readily what he referred to. Of course there survive few other seventh-century attestations to specific water mills, but this is due not necessarily to a decline in the practice of water milling but to the fact that there survive few written attestations to anything from that crucial time. Indeed, the diffusion of all milling technologies in late antique Italy deprives their 'advent' in Carolingian Europe of any spectacular suddenness.

In sum, what appeared to Bloch as an explosion in the charter evidence was most probably only a mirage. Beginning in the mid-700s, the survival-rate of detailed agricultural contracts written in Europe and especially in Italy improved. Agricultural contracts, unlike poems, had every reason to list mills meticulously, with the result that references to water mills in the early medieval documentation are frequent. One side effect of the better survival rate after about 750 of this minute, precise type of document is thus the apparent explosion in the construction of hydraulic mills around that time. Bloch, whose conception of the spread of this technology was quite teleological and rested on assumptions about the ineluctable progress of human ingenuity, saw a dramatic acceleration in the diffusion of water mills in Europe where there may have been none. He minimized the acceptance of hydraulic milling in the ancient world not only because he did not have available the archaeological information unearthed since 1935, but also because he juxtaposed ancient literary sources, full of ambiguous sentiments toward labor, with utilitarian texts that become abundant only for the period after the eighth century C.E.. Other authors, such as the Dutch historian of technology R.J. Forbes, who invoked a supposed 'irrational use of natural forces' by pagan animists, as opposed to a rational Christian use of forces such as flowing water, likewise compared dissimilar texts to reach the conclusion that water milling conquered Europe after Rome was conquered by the barbarians.[15] It is fair to summarize by saying that the sorts of sources that

---

[15] Forbes, 'Power', p. 606; Reynolds, *Stronger*, p. 32 followed Forbes.

survive from and characterize Carolingian Europe and the Roman Empire, and that create the impression of greater use of water milling in the Carolingian world, conceal the gradual, steady, and ancient spread of water milling.

Following Bloch, many students of Europe's mills have considered water milling a significant new element in the landscape of the Carolingian period, or even lightly later.[16] Bloch's dating was central to his interpretation of the dissemination of the water-powered mill. He considered that the lack of manpower, and specifically of slaves to turn millstones, had had an important role in the diffusion of the water mill as a labor-saving device before Carolingian times. But it was the well-documented greed of the Carolingian *potentes*, their desire to establish mills as another means of access to the granaries of the cultivators, which, in Bloch's view, catapulted these machines to success in the ninth and tenth centuries. His position is popular among scholars, who link the growth of large, self-sufficient, lord-owned estates with the dissemination of water milling technologies.[17]

We have already seen that the Roman Empire was far more receptive to the water mill than either Roman or many modern writers allowed. This implies that water mills were popularized in times of abundant slave labor and remained popular during centuries when slavery lost its economic centrality. Availability of servile labor, then, was probably less decisive to the diffusion of hydraulic milling than has been imagined. It is equally difficult to link

---

[16] Bloch, 'Technique et évolution sociale', pp.837-8. The 'boom' in water milling is variously dated within Bloch's chronological framework: to the ninth century (P. Toubert, 'La part du grand domaine dans le décollage économique de l'Occident', in *La croissance agricole du haut moyen âge*, Auch 1990, 68; G. Fourquin, 'Le premier moyen âge', in *Histoire de la France rurale*, ed. G. Duby, Paris, 1975, .362; to the tenth (Parain, 'Rapports', p.64); to the eleventh (P. Benoit, M. Wabont, 'Mittelalterliche Wasserversorgung in Frankreich', in *Geschichte der Wasserversorgung*, v. 4, Mainz, 1991, 192; A. Guillerme, 'Les moulins hydrauliques urbains', *Milieux*, 1 [1980] 45); to the ninth-eleventh centuries (P. Bonassie, 'The Survival and Extinction of the Slave System in the Early Medieval West [Fourth to Eleventh Centuries]', in his *From Slavery to Feudalism in South-Western Europe*, Cambridge, 1991, 39); to the eleventh century (Forbes, 'Power', pp.608-9; J. Langdon, 'Water Mills and Windmills in the West Midlands,' *Journal of Economic History* 44 [1991], 424); to the twelfth century (Reynolds, *Stronger*, p.53; Gille *Histoire*, p. 456). For Italy G. Barni, G. Fasoli, *L'Italia nell'alto medioevo*, Turin, 1971, 623, suggested the 900s. Evidence of water milling in the Jutland and Ireland long before any of these dates (Forbes, 'Power', p. 594; Guillerme, *Le temps*, p.93) is often overlooked.

[17] Eg. J. Le Goff, *La civilisation de l'Occident médiéval*, Paris, 1964, 374; G. Duby, *L'économie rural et la vie des campagnes dans l'Occident médiéval*, Paris, 1962, 73; and *Guerriers et paysans, VIIe-XIIe siècle*, Paris, 1973, 72-3; Dockès, 'Formes et diffusion', pp. 136-8, 141-2, 148-54. Bonassie, 'The Survival', pp. 39-57 added nuance to this picture, showing exceptions in south Europe. Comet, *Le paysan*, pp. 390-96 surveys the French Marxist historiography.

evidence of demographic decline to enthusiasm for labor-saving devices like water mills.[18] Eighth-, ninth-, and tenth-century Italy, in fact, as the extant documents make clear, witnessed wide use of water power but did not experience the labor shortage or population decline it had, for instance, in the sixth or seventh centuries.[19] Rather than being the outcome of population decline, the willingness of early medieval people in Italy to adopt and use hydraulic milling after about 800 may have depended on the period's agricultural vitality and on rural demographic growth.

In her classic essay on *The Conditions for Agricultural Growth*, Esther Boserup showed that growing population levels stimulate technological change in agrarian societies. As a medieval Mediterranean example of this process, it is interesting to consider the spread of hydraulic milling in al-Andalus. In Arab Iberia the success of this technology depended on growing population levels.[20] Thus it is at least possible that Italian water mills were built not to replace missing laborers' muscles, but for completely opposite reasons, for example to grind overabundant grain or to permit the dedication of more energy to land clearances. Duby, in fact, correlates the extension of arable land and grain consumption, signs of population expansion, to the erection and utilization of mills after C.E. 1000.[21]

The important question of why early medieval people built water mills is thus a complicated one that cannot be explained by demographics alone. In fact Bloch advanced another circumstance, that water milling suited the interests of rapacious lords, to explain the explosive success of hydraulic milling. The records, almost all of which come from large estates, seem to bear this out. Certainly most of the more than 5,600 mills listed in the English *Domesday Book* of 1086 were on lords' estates. In Francia too, manorial lords

---

[18]   Along with Bloch, ('Technique', pp. 837-8), Reynolds, *Stronger*, pp. 34, 41; Forbes, 'Power', p. 590, 601, 605; Parain, 'Rapports', pp. 60-62 all saw labor shortage as the key to water power's success. Even a student of monastic architecture, W. Horn, 'Water Power and the Plan of St. Gall', *Journal of Medieval History* 1 (1975) 246, ascribed importance to lack of slaves.

[19]   V. Fumagalli, 'Note sui disboscamenti nella pianura padana in epoca carolingia', *Rivista di storia dell'agricoltura* 7 (1967) 141-4; A. Castagnetti, 'La pianura veronese nel medioevo', in *Una città*, ed. Borelli, 49, 121 exemplify this period of growth.

[20]   E. Boserup, *The Conditions for Agricultural Growth*, Chicago, 1965. Al-Andalus: V. Lagardère, 'Moulins d'Occident musulman au moyen âge', *Al-Qantara* 12 (1991) 69, 71; similar processes have been proposed for medieval Bologna by A. Pini, 'Energia e industria tra Sàvena e Reno: i mulini idraulici bolognesi tra XI e XV secolo', in *Tecnica e società nell'Italia dei secoli XII-XVI*, Pistoia, 1981, 9. See also Comet, *Le paysan*, p. 395.

[21]   Duby, *Guerriers et paysans*, p. 212, though he is dealing with the eleventh century. Also Guillerme, *Le temps*, p. 96.

built water mills in the course of the tenth and eleventh centuries.[22] The ancillary question of why magnates chose expensive and complicated vertical-wheeled machines remains as baffling as the question of the dissemination of water milling in general: neither commercially useful 'efficiencies', nor charitable desire to satisfy serfs' taste for finer flour, nor even lack of water courses suitable for simpler machines are full explanations for the construction of geared, vertical mills on the manor.

The desire to save labor was not acute on Carolingian manors.[23] Bloch, who thought that the manors' serfs were a radically different, more empowered, labor force than Roman slaves, knew this. He sensibly suggested that lords did not build mills to save and redirect serfs' work.[24] Rather, mills were another method of extracting grain from primary producers on the manor. At the mill, lords drew another portion from the peasants' produce in payment for the service of grinding. But water-driven mills represented a very roundabout and complicated means of obtaining extra rents from the tillers of the soil, when simple 'sweat' mills would have done that job as effectively. Hence, the famous greed of the Carolingian magnates is unlikely to have dictated the choice of hydraulic mills.

Leaving these considerable perplexities aside, we should note that manorial lords had much less to do with hydraulic milling in the Italian peninsula than they did north of the Alps. Carolingian capitularies reveal that the powerful landlords in Italy did attempt to impose new exactions on cultivators, though this was never easy and without friction.[25] Enforcing the obligation to use a mill was arduous on the better-run manors of late medieval England, and must have been close to impossible in early medieval Italy, where, indeed, the obligation was never enunciated even in theory. These

---

[22] On Domesday and England, Langdon, 'Water-mills and Windmills', pp. 431-9. Francia: Dockès, 'Formes et diffusion', pp. 132-4. Led by Bloch, Dockès opines (p. 148) that lords imposed use of these mills on unwilling peasants, but acknowledges (pp. 142-6) that enforcement could be problematic. Another proponent of the 'lordly imposition theory' is Guillerme, 'Les moulins hydrauliques', pp. 45-8. See Holt, *The Mills*, pp. 36-52, for a sensible discussion of mill-use enforcement.

[23] Palladius, *Opus Agriculturae* 1.41, ed. R. Rodgers, Leipzig, 1975, p. 75, hinted that water mills saved labor on fourth-century estates, but these were more market-oriented than their Carolingian descendants, whose polyptychs are unconcerned with sparing labor. Though the capitularies discuss magnates who removed the Carolingians' laborers to allodial lands, this does not indicate that the magnates sought to save labor. They sought to accumulate it. The quantity of manpower and not efficiency of work was their concern.

[24] Bloch, 'Les "inventions"', p. 641.

[25] Magnates forced rustics to give up their property and perform extra labor: *MGH Legum Sectio II* 1, ed. A. Boretius, V. Krause, Hannover, 1883, nos. 72.5, p. 163; 93.6, p. 187; 95.13, p. 201 are some examples, easily multipliable.

difficulties with enforcement may explain why in Italy the establishment of machines relying on water technology often sprang from the schemes of urban entrepreneurs or even of peasants who built and managed water-driven mills, not from lords who enforced the use of the mill on unwilling dependent populations.[26] In these cases, peasants and town dwellers demonstrated a fondness and appreciation for water mills. They required no violent inducements to use them.

Why this appreciation of hydraulic mills became so deeply rooted even among subaltern groups requires clarification. Time and toil saved might be proposed as the reasons behind people's willingness to build and use water-driven mills. Even the Augustan epigrammist who first celebrated water mills considered that, once the water nymphs had assumed the duty of grinding, mortal women could toil less and sleep more.[27] Grain milled in the household was indeed ground by hand primarily by women, who were responsible for most of the menial tasks of cooking.[28] That Roman criminals were sometimes condemned to perform this type of work reflects its unpleasant, physically taxing and status-lowering nature. (This, incidentally, explains the heroic willingness of early medieval holy men to mash their own grain, a sign of unmanly humility).[29] In some areas women remained in control of the food-related task of grinding at the water-driven mill. This actually created a new type of work for them, as water mills, unlike other muscle-powered grinders, were seldom located at convenient points close to houses, but arose instead along suitable streams. In such cases, hydraulic mills did not save labor so much as change its nature.

---

[26] Toubert, 'La part', pp. 69-73 vigorously affirms the importance of manorial lords in mills' construction. In this he follows Bloch, 'Avènement et conquêtes', p. 541 who described the 'double constraint' behind the mills' diffusion: population decline and manorial oppression.

[27] *Anthologia Graeca Epigrammatum* 9.418, ed. H. Stadtmueller, v. 3, Leipzig, 1906, 402-3 (good commentary on this text is in E. Curwen, 'The Problem of Early Water-mills', *Antiquity*, 18 [1944] 134-5). Duby, *Guerriers et paysans*, p. 73 also suggested medieval peasants liked water mills because they saved time and effort.

[28] Bloch, 'Avènement et conquêtes', pp. 551-2 remarked on Germanic chieftains' need of female slaves to grind grain. On female domestic labor: E. Ennen, *Frauen im Mittelalter*, Munich, 1994, 88; D. Herlihy, *Opera Muliebra. Women and Work in Medieval Europe*, Philadelphia, 1990, 34-6. For Italy: B. Andreolli, 'Tra podere e gineceo. Il lavoro delle donne nelle grandi aziende agrarie dell'alto medioevo', p. 31; and P. Galetti, 'La donna contadina. Figure femminili nei contratti agrari dell'alto medioevo', p. 49, both in *Donne e lavoro nell'Italia medievale*, ed. M. Muzzarelli, P. Galetti, B. Andreolli, Turin, 1991.

[29] *Codex Theodosianus* 9.40.3 (C.E. 319), p. 501; 9.40.5-7 (C.E. 364), p. 502. Constantius of Lyon, *Vita Sancti Germani Episcopi* 1.3, ed. R. Borius, Paris, 1965, 126. Germanus resembled Plautus, who spent his days grinding meal for pay, but rose above these circumstances (he wrote fables by night).

To grind grain at a mill, however, was likely to involve males, who transported the grain, supervised its grinding, and brought home the flour. The goings-on inside early medieval mills seldom emerge from the charters, but when we are offered a glimpse of a mill's interior it is full of men. Pope Gregory I's *Dialogues* relate a miracle that took place in a mill, involving only males.[30] In cases where men became involved, water mills transferred the burden of preparing flour across gender lines and also added some new chores. In rural areas where men assumed the responsibility of grinding grain, peasant women gained extra time to spend performing other work, while peasant males lost time in the fields. Similar transfers of work loads occurred in urban contexts, where males took over the burden of machine grinding. Therefore, just as the willingness of lords to build water-driven mills—an expensive affair—is not immediately logical, the willingness of subalterns to build and to use them, though they paid multures and risked fraud, is unlikely to have resulted simply from a desire to 'save labor'. The water mill improved the 'quality' of the work, rendering the preparation of flour less exhausting for those who mashed grain at home, and perhaps also less humiliating. However, the water mill did not automatically diminish the quantity of work, or the time dedicated to grinding, for the operations it required were numerous and cumbersome.[31]

In early medieval Italy the dissemination of water milling was a complex phenomenon linked only in part to the cupidity of manorial lords or the enthusiasm of peasants to free up additional time in which to till the soil (a dubious concept in any case). Powerful Italian lords, urban entrepreneurs, and even peasants built these machines because they found them useful. Their understanding of utility, of course, varied. Profits derived from the toll on milled grain was a consideration for everyone, perhaps for peasants too. At Amalfi, urban capitalists consciously invested money in mills hoping to obtain long term returns.[32] For cultivators, whatever the exact terms of the equation were, water mills changed the way in which their households allocated labor resources, and many found the mill-determined equilibrium advantageous. At the other end of the social scale, for the elites in early medieval Italian society, prestige was also attractive; water-driven machines, the grander the better,

---

[30] *Dialogi de Vita* 3.37, ed. A. de Vogüé, P. Antin, Paris, 1979, v. 2, p. 412. This is a 'sweat' or muscle-powered oil mill.
[31] Both Dockès, 'Formes et diffusion', pp. 135-50; and Holt, *The Mills*, pp. 36-52, 69, treat the subject of gender, time saved, and water power well, but reach opposite conclusions on the way the technology spread.
[32] Del Treppo, *Amalfi medioevale*, pp. 45, 48; P. Skinner, *Family Power in Southern Italy*, Cambridge, 1995, 259-61.

may have added glamour to the owners' reputation and perhaps associated the owners with public authority. Lordly mills could also symbolize the lords' authority over the people on the land or stand for the unity of the lineage whose various members kept part-shares of the mill long after estates had been divided among heirs.[33] Certainly the willingness to build vertical-wheeled water mills cannot always be explained as a rational economic calculation, and various cultural constraints obviously drove the dissemination of grinding technologies. Among such constraints was the desire for a finer, whiter flour, better suited to the confection of breads. Neither hand querns nor 'sweat' mills could match the even, regulated grinding of water mills, whose flour was smooth, pale, and ideal for bread. In early medieval Italy, bread retained all of the prestige it had held in Antiquity, and it perhaps remained central enough to the diet to override considerations of cost in labor or capital.[34] Peasants who built water mills or willingly used manorial mills did so to obtain a food they valued highly.[35] In addition to all these considerations, we should not lose sight of the fact that water mills, especially those with horizontal wheels, were normal and familiar in the early medieval peninsula. The wide distribution of hydraulic milling in postclassical Italy thus grew from the successes of this technology in the ancient Mediterranean.[36]

Though it is clear that the advent and conquests of the water mill in early medieval Italy arose from a variety of factors, a fuller comprehension of this process requires a chronologically nuanced, regional analysis of the use of this technology. Patterns within the diffusion process are difficult to reconstruct for the darker Dark Ages. Nevertheless, it is reasonable to surmise

---

[33] Duby, *Guerriers et paysans*, p. 71, suggested mills were prized symbols of *romanitas* among Germanic aristocrats. Skinner, *Family Power*, pp. 75-8, 264-5 connects mill-holding to elite status in tenth-century Campania, and stresses the 'public' associations of grain supply and water control.

[34] Water mills could also make mash suitable for porridges, naturally. On bread consumption, M. Montanari, *L'alimentazione contadina nell'alto medioevo*, Naples, 1979, 193-4, 211-8, who is reductionist. Holt, *The Mills*, p. 10 accuses porridge-consumption of slowing the spread of hydraulic milling.

[35] Income levels from the Carolingian estate at Annapes prove more grain was ground at the mill than was grown on the manor, suggesting that willing outsiders may have brought grain to the water mill for grinding: Duby, *Guerriers et paysans*, p. 73. Similar patterns exist in the Brescian S. Giulia's polyptych.

[36] Large, state-run vertical-wheeled mills became less common, without disappearing, but in rural contexts the continuity of the horizontal wheel's technology was strong. It is conceivable that the many water mills mentioned in the postclassical charters arose on the sites of rural Roman installations. Roman hydraulic knowledge would determine the selection of the best sites, and continuous use of these might explain why so many Roman horizontal mills 'disappeared'.

that in the earlier period, before C.E. 700, as in the later one, up to C.E. 1000, hydrological and other environmental conditions affected how water mills survived or spread. Social structures and cultural norms had an equal impact. These constraints dictated considerable regional variations in the dissemination of this technology.

Thus water milling developed little around Ravenna, whose flat and marshy floodplain, full of sluggish water courses, hampered the use of this technology. The persistence of Byzantine administration until 751, and the resilience thereafter of Roman law, with its many restrictions on the use of navigable water courses by private individuals, also militated against the installation of water mills around Ravenna.[37] Hydraulic milling developed only slightly more in the region of Rome, where geography and hydrological conditions were favorable but legal traditions were less so.[38] Yet the Sabine hills northeast of Rome sheltered several hydraulic mills from the eighth century forward. These machines belonged to the great abbey at Farfa or to the bishops of Rieti, and lay people appear to have had a negligible role in their construction, particularly in the ninth and tenth centuries.[39] After about 850, therefore, the Sabina might reflect the patterns of mill dissemination by powerful, wealthy landowners which scholars have detected in Francia. The new patterns of human settlement on hilltops in the Sabina, dating to the late Carolingian period and thus contemporary with these water mills, may also be related to the mills' construction. The relocation of settlements, called *incastellamento*, was often carried out under lordly direction, so the erection of these Sabine mills could be part of an aristocratic strategy of resource management.[40] But the situation was again different at Salerno in the south of Italy and in the central Po valley, where milling by water was a commonplace as far back as the documentation reaches. In these dissimilar places, stream regimens were suited to hydraulic mills. Local magnates, together with urban professionals, churches, and even some smallholders (at least until the mid-

---

[37] P. Squatriti, *Water and Society in Early Medieval Italy*, University of Virginia, 1990 PhD, ch. 4.

[38] On water mills near Rome in the ninth century, J. Raspi Serra, C. Laganara Fabiano, *Economia e territorio. Il Patrimonium Beati Petri nella Tuscia*, Naples, 1987, 193, 241.

[39] Sabine mills: *Regesto di Farfa* 2, ed. I. Giorgi, U. Balzani, Rome, 1879, 251:208 (possibly in Ascoli); 290:245; 293:247, all of mid-ninth century date. P. Toubert, *Les structures du Latium médiéval*, v. 1, Rome, 1974, 460, notes the existence of some peasant-owned mills. See also B. Condorelli, 'La molitura ad acqua nella valle del torrente Farfa', in *Atti del 9° congresso internazionale di studi sull'alto medioevo*, Spoleto, 1983, 838-40.

[40] A brief survey of *incastellamento*: K. Randsborg, *The First Millennium AD in Europe and the Mediterranean*, Cambridge, 1991, 65-71. Tuscan and Molisan *incastellamento* is difficult to relate to water mills.

800s) built and managed them. In both southern Italy and in Lombardy, legal traditions encouraged the appropriation of water resources for milling—a trend reinforced in Lombardy by the later collapse of the state. By contrast, in northwestern Tuscany from the eighth century until at least the eleventh, simple Lucchese cultivators and ecclesiastical foundations exploited the water courses to grind grain and, in the case of the churches, to provide incomes.[41] The plain of Lucca had a sufficient gradient that water flowed reliably, and the longevity of comital authority there prevented the usurpation of mills by the *potentes*.

The history of the distribution of water milling technologies in Italy in the early Middle Ages is therefore a regional history determined by the twin constraints of ecology and cultural expectations. It is also an asymmetrical history. Even as paddle wheels of diverse types slapped the waters of some streams, the waters of other regions flowed undisturbed by such human impositions. And everywhere several methods of grinding grain, with and without water, coexisted. For, as Marc Bloch knew, the advent and conquests of the water mill, like the advance of any technology, was never a total victory, despite its very long campaign.

---

[41] Squatriti, *Water and Society*, ch. 2, 4-5; also, P. Squatriti, 'Water, Nature, and Culture in Early Medieval Lucca', *Early Medieval Europe* 4 (1995) 30-36.

7

Mechanization and the Medieval English Economy

Richard Holt

Domesday Book, that unique survey of the principal economic assets of an entire medieval country, records some 6,082 mills in England in 1086. Even that impressive total is not a complete one, as the northern counties were either poorly recorded or not at all; in all probability, eleventh-century England was served by 6,500 mills or more. Machinery driven by a natural source of power was therefore not a rarity, although even a crude estimate of its importance relative to other power sources puts it into perspective: over the whole country there were 75 mills for every 1,000 plough teams, each of a notional eight oxen, so that each mill was matched by approximately 100 working oxen.[1] Taking into account the unknown number of horses and the two million or more labouring people, it is clear that this society, for all its familiarity with mechanization, was overwhelmingly dependent for motive power on the muscles of people and animals. Four centuries later, at the end of the Middle Ages, that was still the case.

For each mill, Domesday Book records its owner and its rental value. What more can be deduced? It is an assumption, although a safe one, that these were all watermills. The mapping of Domesday mills county by county by H.C. Darby and his co-authors in their extended study of the geography of Domesday England demonstrates how closely the distribution of mills in every part of England was related to the availability of suitable watercourses.[2] Moreover, there is meaning in the all-embracing terminology of 1086. The clerks of later centuries nearly always distinguished between windmills and

---

[1] H.C. Darby, *Domesday England*, Cambridge, 1977, 361.
[2] H.C. Darby, *The Domesday Geography of Eastern England*, Cambridge, 1952; H.C. Darby and I.B. Terrett, *The Domesday Geography of Midland England*, Cambridge, 1954; H.C. Darby and I.S. Maxwell, *The Domesday Geography of Northern England*, Cambridge, 1962; H.C. Darby and E.M.J. Campbell, *The Domesday Geography of South-East England*, Cambridge, 1962; H.C. Darby and R.W. Finn, *The Domesday Geography of South-West England*, Cambridge, 1967.

watermills; their predecessors could use 'mill' as a universal term because it had an unambiguous meaning, all manorial mills being watermills. In the same way, it is certain that these recorded mills were all cornmills, grinding a variety of grains into meal for making bread or porridge, or grinding the malted grains used in brewing ale. From the following centuries there are examples of the application of the mill mechanism to a range of industrial processes, but medieval clerks always carefully differentiated such machines by name from ordinary cornmills. Again, the lack of differentiation in Domesday Book demonstrates that the eleventh-century clerks recognized the ubiquitous watermill as having only a single function.

Although our documentation effectively begins with Domesday Book, the history of the English mill goes back much further, and the scatter of references to mills in Anglo-Saxon charters dating back to 762 may be referring to a machine in common use.[3] Only two Anglo-Saxon mills have been excavated: a mill at Old Windsor on the Thames, dated by dendrochronology to 690, and a ninth-century mill excavated at Tamworth in the Midlands.[4] But contemporary Ireland made wide use of the mill, and modern peat digging has so far brought to light more than thirty from that period. Dendrochronology provides dates consistent with the sixth- and seventh-century references to mills in Irish law codes and saints' lives, the earliest mill discovered to date having been built of timbers felled in 630.[5] It is not certain how the Irish experience related to that of other parts of Europe which have not produced evidence in such quantity; nevertheless, there is a growing appreciation that the watermill, which historians now realise had become widely accepted during the Roman period, is likely to have been a common sight throughout most of Europe in the early Middle Ages.[6]

[3] P.H. Sawyer, ed., *Anglo-Saxon Charters*, London, 1968, 25; W. de Gray Birch, ed., *Cartularium Saxonicum*, London, 1885-99, 191; D. Hill, *An Atlas of Anglo-Saxon England*, Oxford, 1981, 114.

[4] P. Rahtz and R. Meeson, *An Anglo-Saxon Watermill at Tamworth*, Council for British Archaeology Research Report 83, London, 1992, 156 and passim.

[5] C. Rynne, 'The Introduction of the Vertical Watermill into Ireland: Some Recent Archaeological Evidence', *Medieval Archaeology* 33 (1989) 21-31; M. Baillie, 'Dendrochronology: The Irish View', *Current Archaeology* 73 (August 1980) 61-3; P. Rahtz and D. Bullough, 'The Parts of an Anglo-Saxon Mill', in P. Clemoes, ed., *Anglo-Saxon England 6*, Cambridge, 1977, 15-37.

[6] Ö. Wikander, *Exploitation of Water-Power of Technological Stagnation? A Reappraisal of the Productive Forces in the Roman Empire*, Lund, 1984; Ö. Wikander, 'Archaeological Evidence for Early Watermills: an Interim Report', *History of Technology* 10 (1985) 151-79; K. Greene, 'Perspectives on Roman Technology', *Oxford Journal of Archaeology* 9 (1990) 209-19.

The point is a crucial one, as so much has in the past been built upon the erroneous belief that the watermill, although an invention inherited from Antiquity, only came into significant use during the High Middle Ages. That view was the cornerstone of Lynn White Jr.'s theory that medieval people had a special interest in powered machinery; the cornmill, White believed, was only the first of the many applications of waterpower and later windpower by means of which the people of the Middle Ages strove to transform their world.[7] Among medieval historians, the theory has never found favour beyond a narrow circle of White's disciples; it has, though, been widely propagated among non-specialists and as a consequence has had a significant impact. But how far in reality did medieval people look to powered machinery as a useful aid to production? What really was the economic impact of the machines they built?

Answers to such questions become more feasible with the great expansion of record-keeping which in England came with the thirteenth century. A range of documentation from town and country, both publicly and privately commissioned, provides a wide scatter of information on medieval agriculture and industrial production to set beside the valuable and still growing body of archaeological evidence. The greatest volume of material relating to mechanization is to be found in the numerous series of estate accounts, detailing *inter alia* the working of manorial mills and thus constituting our prime source of evidence for their construction, working and profitability. The fact that virtually all were built to grind corn reflects the dominance of bread in the medieval diet, and the constant demand for flour as a commodity of prime importance: the cornmill was, and for centuries would remain, the most profitable and far and away the most frequent application of natural power. The variety of factors, therefore, compelling and constraining the application of waterpower and wind power, but also muscle-power as well, to the production of flour must be central to our appreciation of the circumstances of medieval mechanization.

CORNMILLS

In manorial records, mills appear generally as demesne assets that generated revenue and were a constant liability; each annual account, therefore, will record the income from any mills the lord of the manor owned, together with whatever amounts had been spent on repairs and wages. The purpose of the

---

[7] L. White, Jr., *Medieval Technology and Social Change*, Oxford, 1962.

account, as it emerged in twelfth-century England, was to prevent or at least restrain cheating by the manorial officials, and the auditors demanded precise details of both income and expenditure. In many accounts, as a consequence, spending on mills was justified to the last penny with every nail enumerated and every foot of timber or every replacement wooden tooth for the cogwheels accounted for. Whilst such information cannot easily be translated into the description of a mechanism, nevertheless certain fundamentals of mill design are clear. That these mills used cogwheels and gearing is a sign that they all employed vertical waterwheels; the less elaborate horizontal-wheeled mill of the earlier Middle Ages which remained in wide use in parts of Europe had no place on the thirteenth-century English manor. Mills lacked the power to drive more than a single set of millstones; if greater capacity was needed, a second mechanism was required—and often installed within the same building—to drive the additional stones. That lack of power was the result of the mainly timber construction of these mills, the only moving part of iron being the drive shaft to the millstone; in particular, the size and thus power of the waterwheel was severely restricted until the availability of cast iron in the eighteenth century made it feasible to construct waterwheels that were larger, more robust, and thus much more powerful.[8] Needless to say, the serious limitations of the medieval mill mechanism must have severely restricted attempts to adapt it to industrial processes other than cornmilling.

Yet despite its limitations, the profitability of the manorial mill—at any rate before the fifteenth century—is not in doubt. It was reinforced by the practice, valid in customary law, of compelling the tenants of the manor to use only their lord's mill. The toll paid for grinding was generally around one-twenty-fourth of the grain, although in the north of England it rose to an exorbitant one-thirteenth. Any attempt to use a mill other than the lord's, or to mill at home with a handmill, would—if detected—lead to a fine in the manorial court.[9] It is this coercive aspect of the mill which in the past has most interested historians: for Marc Bloch, it was the essential factor in the mill's profitability and thus the crucial reason for its ubiquity. In an influential article, he advanced the view that this example of mechanization had come about as a by-product of seigneurial extraction; it was not any capacity to grind corn efficiently and cheaply which made mills attractive to the lords who built and maintained them but their usefulness as a tool of social and economic control. By means of the mill, lords could centralize and regulate the basic activity of grinding corn and so draw a revenue which would

---

[8] Richard Holt, *The Mills of Medieval England*, Oxford, 1988, 122-32.
[9] Ibid., pp. 36-53, especially 49-51.

otherwise be beyond their grasp. To support his contention, Bloch quoted a series of epic struggles between lords and their tenants: in all of them the tenants were apparently implacably opposed to this machine which transformed the production of flour from a domestic activity into an industrial process beyond their control.[10]

Bloch's deep understanding of the dynamics of medieval society led him to observe and appreciate a social dimension to the mill which it certainly possessed, although not quite as he saw it. He was convinced that the watermill was essentially a novelty even in the tenth and eleventh centuries: writing in 1935, he could not anticipate the growth in archaeological evidence for so many mills from the Roman period and from the early Middle Ages. Despite what he believed, medieval lords and their tenants were not quarreling over a new aristocratic levy, though perhaps they were in contention over an increased rate of extraction; Bloch's evidence for particularly bitter conflicts came largely from northern France and other regions where there is abundant evidence for a maximisation of profits from the rights of lordship—*seigneurie banale*--which included milling, in compensation for falling rent revenues.[11] But disputes over milling were often not so simple as Bloch thought; at St Albans, for instance, what he saw as a passionate refusal to use the abbot's mill was in fact a deliberate act of defiance—in effect a strike—in a long running struggle between prosperous townsmen and a grasping lord who refused to grant them the urban liberties enjoyed by burgesses elsewhere. Their insistence on using their own handmills says nothing, in fact, about the preferences of the townspeople, just as the abbot's violent reaction was prompted by more than just a desire to protect his milling revenues. In normal years the abbot of St Albans, like other lords, was prepared—for a moderate fee—to license either handmills or more often a general exemption from use of the manorial mill.[12]

Hand querns were commonly used, often legally, and especially for lighter tasks such as bruising oats or barley for porridge, or crushing malt for brewing. But the traditional view that peasants, and indeed townspeople, retained a preference for milling by hand probably underestimates their aversion to this irksome task. The view's weakness is that it ignores the

---

[10] M. Bloch, 'Avènement et conquêtes du moulin à eau', *Annales Économies, Sociétés, Civilisations* 7 (1935) 538-63, trans. J.E. Anderson in M. Bloch, *Land and Work in Medieval Europe*, London, 1967, 136-68.

[11] G. Duby, *Rural Economy and Country Life in the Medieval West*, trans. C. Postan, London, 1968, 252.

[12] Holt, *Mills of Medieval England*, pp. 40-41, 44-5.

scattered but substantial body of evidence indicating that the real alternative to the manorial mill was the private horsemill. Animal-powered mills were known to the Romans, and might have been in widespread use during the early Middle Ages, long before they were first recorded during the twelfth century.[13] Domesday Book makes no reference to them, but that does not mean we can safely assume they did not exist; private horsemills serving the needs of a single household were not revenue-producing manorial assets, and so would have been ignored by the commissioners. Nor is the lack of excavated examples significant as a small, semi-portable mechanism of this sort would be difficult or impossible to detect archaeologically. The excavator of the palace of the Wessex kings at Cheddar, a prominent archaeologist deservedly respected for the quality of his expertise, interpreted a tenth-century circular building first as a horsemill and later, without real justification, as a hen house—as good an illustration as any of how hopeless a task it often is to assign any function to excavated buildings.[14]

Even in later centuries, horsemills tended to remain below the documentary threshold. Manorial horsemills were extremely rare, although this source provides the only real clues to their performance and the potential of the technology. On the evidence of the annual accounts from Glastonbury Abbey's manor of Westonzoyland in Somerset, the horsemill there was a complex machine with a geared mechanism. Built before 1274 to replace an earlier windmill, until at least 1335 it continued to command an annual rent of £1 10s. or more, three-quarters of the £2 which the abbey usually received from its windmills in the district.[15] Presumably the domestic horsemills, to which we have most references, were seldom as elaborate or as powerful as that. The bishop of Ely's manor of Wisbech in Cambridgeshire had no mill when it was surveyed in 1251, and the surveyor expressed his assumption that not only the bakers of the manor but also the agricultural tenants would own their own horsemills.[16] The nature of the machine—relatively unsophisticated and relatively compact, and with its manageable power source—made it particularly suited to urban use; in the larger towns, at least, a milling monopoly rarely applied, and there are references to bakers and brewers keeping horsemills. One example comes in the lease of a house and brewery in

[13] Ibid., p. 17.

[14] P.A. Rahtz, 'The Saxon and Medieval Palaces at Cheddar', *Medieval Archaeology* 6-7 (1962-3) 53-66; P.A. Rahtz, *The Saxon and Medieval Palaces at Cheddar: Excavations 1960-62*, British Archaeological Reports, British ser., 65, London, 1979

[15] Longleat House, MSS 11,244; 10,766; 10,656; 10,632.

[16] British Library, Cotton MS Claud. Cxi, fols. 74r, 76r.

Gloucester in 1429, equipped with boilers and a great brewing vat, and a horsemill valued at ten shillings. That was three times as much as the vat was worth, and the whole property was to be rented for £1 10s. a year, putting the value of this mill into further perspective.[17] Another, this time famous, example is that of Margery Kempe. This pious pilgrim and autobiographer was the wife of a wealthy burgess of Lynn in Norfolk, and in the 1390s had her own business interests. First she became one of the leading brewers of the town, but in time her enterprise failed: because of her sinfulness, her ale repeatedly failed to ferment. So then she turned to commercial milling, with the horsemill she already owned. But her sinfulness again brought disaster, when neither of her two horses would take its turn at working the mill, and the man she employed resigned in disgust.[18]

The horsemill was doubtless a machine best suited to small-scale enterprise, and further examples come from the fifteenth century when in many villages social changes and a decrease in population made large-scale milling uneconomic for many lords to pursue.[19] Large-scale milling was reserved to the watermill, and after the twelfth century to the windmill as well. The most successful of wind-powered machines—the sailing ship—was already an ancient device, but the European windmill was a new invention of the medieval West, appearing first in northwest Europe and in England not long before 1185. The earliest securely dated references come from that year, whilst not a single windmill reference supposedly earlier than the 1180s stands up to critical examination.[20]

The medieval post mill, its whole superstructure turning on a single great post to keep the sails facing the wind, was a great medieval achievement. It was the most characteristic contribution of the age to the technology of exploiting natural power, and by far the most important. Not that we should exaggerate that importance, however, as Lynn White did. He insisted that the rapid spread of the windmill was 'fundamental to our understanding of the

---

[17] Public Record Office, C115/K2/6682, fol. 200r.

[18] S.B. Meech and H.E. Allen, eds., *The Book of Margery Kempe*, Early English Text Society, O.S. 212, London, 1940, 9-11.

[19] Holt, *Mills of Medieval England*, pp. 166-70.

[20] Ibid., pp. 20-21; A-M. Bautier, 'Les plus anciennes mentions de moulins hydrauliques industriels et de moulins à vent', *Bulletin Philologique et Historique* 2 (1960) 567-626. The case advanced by Edward Kealey, *Harvesting the Air: Windmill Pioneers in Twelfth-Century England*, Woodbridge, 1987, that the windmill was in use well before 1185, is unconvincing: see Richard Holt, 'Milling Technology in the Middle Ages: the Direction of Recent Research', *Industrial Archaeology Review* 13 (1990) 50-8 on p. 54.

technological dynamism of that era'.[21] Following Bloch in his conviction that the watermill had only recently spread through western Europe, White envisaged a period when people eagerly sought ways to harness natural power. He mistakenly believed that watermills driven by the tides were an innovation of the eleventh century, and the subsequent invention of the windmill seemed to confirm his thesis.[22] Yet, seen in detail, the spread of the windmill was less impressive than he imagined. First recorded in both England and Normandy during the 1180s, most early examples come from the better English evidence: there are at least 23 English windmills safely dated to between 1185 and 1200, all on the less hilly and often drier eastern side of the country. The following century saw knowledge of the windmill spread out from this area of origin to reach most of western Europe. A close analysis of the English evidence, however, shows that it became established rather slowly; even in eastern England where the windmill's impact was greatest, the main period of windmill building came at least a generation after its initial appearance, between 1220 and 1250. The impression we receive is that at first no more than a scatter of landowners were willing to invest in the new technology, with most waiting for as long as fifty years before they would invest in it.[23]

Typical of those displaying such a notable degree of caution were the bishops of Ely, in other respects great improving landlords who were willing to invest heavily in draining vast areas of fenland. Their estates were poorly served by waterpower, and like Ely itself were all in East Anglia, the region where most of the first windmills were sited. But on the bishopric's fifty manors there were only four windmills in 1222; up till that date the bishops had certainly had no particular enthusiasm for this new power technology. It was during the following thirty years that a determined programme of windmill building began, and a later estate survey from 1251 shows now not four but thirty-two windmills—an impressive eight-fold increase.[24] Other large estates in East Anglia show the same pattern; Ramsey Abbey's estates were in similar country to Ely's, and although in time Ramsey would invest heavily in windmills, there was little sign of this before the middle of the thirteenth century.[25] Certainly these great lords knew about windmills: in fact it had been the abbot of Ramsey and the archdeacon of Ely who back in the 1190s had first referred to the pope the question of whether or not windmills should pay

[21] White, *Medieval Technology and Social Change*, pp. 87-8.
[22] Ibid., p. 85.
[23] Holt, *Mills of Medieval England*, pp. 20-35, 171-5.
[24] British Library, Cotton MS Tib. Bii, fols. 86-241; Cotton MS Claud. Cxi, fols. 25-312.
[25] Holt, *Mills of Medieval England*, p. 24.

tithes.[26] Their reluctance to build windmills themselves, when there were some landowners who evidently found them profitable investments, points to a powerful streak of conservatism quite at odds with the supposed technological dynamism of which White was so sure.

The total of windmills scarcely increased after 1280, and by 1300 the building boom had finished. By this time the distribution pattern of the windmill was firmly established, with by far the greatest concentrations to be found in districts where water resources were poor. In parts of eastern England most mills now were windmills, whereas the watermill massively predominated elsewhere.[27] The copious details about mills taken from manorial accounts confirm that watermills generally commanded significantly higher rents than windmills, or could be operated more profitably; furthermore, windmills were relatively expensive to maintain, requiring one-third of their revenue over a period to be spent on repairs.[28] The superiority of the watermill over the windmill is obvious; the lords who built windmills did so because they lacked water resources and there was no other way in which they could tap into the profits that their more fortunate neighbours already received from milling.

An illustration of that point, and of the circumstances surrounding the short life of one of the earliest windmills in England, comes from the twelfth-century chronicle of Jocelin of Brakelond. Jocelin was a monk of the great abbey of Bury St. Edmunds, which owned and had jurisdiction over a large part of western Suffolk. Around the abbey a wealthy town had grown up, and relationships between the monks and the townsmen whose activities they tried to control were generally uneasy. Jocelin described how in 1191 the formidable Abbot Samson flew into a fearful rage because a cleric called Herbert, the rural dean of Bury, had built a windmill on his own land just outside the town. On being summoned, Herbert protested that he had done nothing illegal, and that the mill was for his own use—on the face of it, an unlikely story. Samson pointed out that it would compete with his own mills, because the burgesses, as free men, could mill where they liked. He reminded Herbert that he had complete authority over the district, meaning that it was his right to decide if a windmill was illegal or not. Thus, whilst the abbot did not have the right to compel the townspeople to use his own mills,

---

[26] M.G. Cheney, 'The Decretal of Pope Celestine III on Tithes of Windmills, JL 17620', *Bulletin of Medieval Canon Law* new ser. 1 (1971) 63-6.

[27] Holt, *Mills of Medieval England*, pp. 25-34, 108-11.

[28] Ibid., pp. 77-8, 86-7.

nevertheless he had the power to enforce a local monopoly simply by ensuring that only his own mills could operate within the area of his jurisdiction. Herbert had no choice but to back down, and straight away gave orders to have his mill dismantled before the abbot's men came to destroy it.[29]

We have now seen a series of factors—some competing—that together conditioned the way in which the windmill had an impact on English society. The notion that the new device was eagerly accepted because it extended the exploitation of natural power is unhistorical; there is no evidence at all that the people of the Middle Ages took such an attitude. Instead, the views of different individuals depended on their circumstances. Some, like Herbert, saw the prospect of profit for themselves, particularly if there was a readily identified local market for a new mill; by contrast, powerful men like Samson—owners of watermills—were concerned to protect their existing interests and saw no benefit to themselves in the windmill. As for the customers of the new machine, Samson clearly believed that the burgesses of Bury, free to mill where they wanted, would welcome some local competition with the abbey's watermills, although such a positive view of the windmill would not necessarily have been echoed by the mass of the people in the countryside. A lord of a manor without a watermill who then built a windmill did so to make a profit, and would require his tenants to use it and to pay the tolls he set. That might have been a less advantageous arrangement for the peasants than whatever arrangements they already had for milling. Another factor in the equation was the windmill's vulnerability to damage, and the greater liability borne by its owners—and ultimately by its customers—to meet the costs of repairs.

The number of English medieval mills continued to increase until the decades around 1300. Precise figures are impossible, but broad samples from a number of regions together indicate that the 6,000 or more watermills of 1086 had become 9,000 or more, and that 3,000 or more windmills had been built. Everything points to at least a doubling of the number of English mills since Domesday Book. That is hardly surprising, given the equivalent increase in the population during the same period. We have no information as to the total capacity of these mills, either in the eleventh century or the fourteenth; whilst the later mills may have been more powerful than some of the eleventh-century mills—particularly if the intervening period had seen the phasing out of small, horizontal-wheeled mills—that cannot be demonstrated. It is

---

[29] H.E. Butler, ed., *The Chronicle of Jocelin of Brakelond*, London, 1949, 59-60.

therefore impossible to tell by how much milling capacity had increased in relation to the population, or indeed whether it had increased significantly at all. In other words, there is no evidence to show that the people of the fourteenth century made greater use of the mill than did their ancestors of nearly three centuries before.[30]

The apparent correlation between the number of mills and the number of people is also to be observed during the long period of population decline which began around the middle of the fourteenth century. Many mills fell into disuse in the aftermath of the Black Death of 1349 and subsequent plague epidemics, and the process of decline was renewed as the fifteenth century opened.[31] It has already been suggested that this period may have seen greater use of the horsemill as large-scale milling declined; such a setback to the aristocracy's control of milling can only have occurred as part of its wider loss of authority over its agricultural tenants, and its consequent crisis of income.[32]

MILLS IN MANUFACTURING INDUSTRY

Before examining the extent of mechanization of industrial processes other than cornmilling, it may be useful to stress that every medieval English windmill was built to grind grain, as was the great majority of watermills. Both the cereal diet of the mass of the people and the poverty that restricted expenditure mainly to foodstuffs[33] ensured that the production of flour would be by far the most important industrial process of the period. The volume of food grains to be processed in virtually every community meant that the level of profit was sufficient to persuade lords to invest in a mill. Yet even the profitable cornmill was not always an ideal investment, and with heavy expenditure on repairs to mechanism or to watercourse, and on costly new millstones[34], many mills worked perilously close to the margin.

This question of profitability has nearly always been ignored by writers on industrial mills. Lynn White believed that the wider use of waterpower was yet another triumph of a medieval enthusiasm for

---

[30] Holt, *Mills of Medieval England*, pp. 115-16.

[31] Ibid., pp. 159-64.

[32] John Hatcher, *Plague, Population and the English Economy 1348-1530*, London, 1977, 35-42.

[33] Christopher Dyer, *Standards of Living in the Later Middle Ages*, Cambridge, 1989, 151-60.

[34] Holt, *Mills of Medieval England*, pp. 86-8, 99-100, 161, 176-7; David Farmer, 'Millstones for Medieval Manors', *Agricultural History Review* 40 (1992) 97-111.

mechanisation, and a handful of other writers have followed that line.[35] But White's technique was simply to note a scatter of references to a range of waterpowered machines, and then to proceed to the conclusion that their economic impact must have been considerable. In fact, in the absence of extended studies of industrial mills, the precise extent of that impact remains unknown. What is certain, though, is that with their relatively small number and limited capacity the impact of industrial mills must have been slight. Analysis of large numbers of English mills shows just how rare the various types of industrial mill were. Sawmills are not to be found, in contrast with the huge numbers of references in manorial accounts to the wages paid to sawyers, sawing timber by hand. Mills for grinding bark for tanning, waterpowered forges used to hammer the blooms produced in the smelting process, and mills for smithing or for grinding and sharpening blades, are all to be found in very small numbers. Only fulling mills are recorded often enough to escape being classed as uncommon, although only in districts where water resources exceeded the requirement for cornmilling.[36]

Fulling is the process in which newly woven woollen cloth is vigorously washed to scour out the natural oils of the wool, and to shrink and felt the fabric. Traditionally it was done by hand, and more usually under foot, in large troughs of water and using various detergent substances. The process of beating and agitating the cloth was one that readily lent itself to mechanisation, using a waterwheel to power a series of rising and falling hammers, and fulling mills were recorded in Normandy in 1087 and at a number of European locations during the following century. The earliest and still the only sustained account of the fulling mill's introduction and impact is Eleanora Carus-Wilson's study, published in 1941, of the English evidence.[37] It had appeared in England by 1185, and thereafter references to it are numerous: Carus-Wilson and others pointed to over 140 English and Welsh fulling mills—not all operating at once—during the period before 1350.[38] Most were in the well watered west and north of the country, away from the

---

[35] For instance, see Terry Reynolds, *Stronger Than a Hundred Men: A History of the Vertical Water Wheel*, Baltimore, 1983, 47-8; Jean Gimpel, *The Medieval Machine*, London, 1977, 1979 ed., 15-40.

[36] Holt, *Mills of Medieval England*, pp. 147-52.

[37] E.M. Carus-Wilson, 'An Industrial Revolution of the Thirteenth Century', *Economic History Review* 11 (1941) 39-60, repr. in E.M. Carus-Wilson, ed., *Essays in Economic History*, v. 1, London, 1954, 41-60, and in E. M. Carus-Wilson, *Medieval Merchant Venturers*, London, 1954, 183-210.

[38] R.V Lennard, 'Early English Fulling Mills: Additional Examples', *Economic History Review* 2nd. ser. 3 (1950-1) 342-3.

cloth industry of the great towns of the south and east, and Carus-Wilson incautiously proposed that the effect of the new machine had been to encourage some relocation of the cloth industry away from the established centres. That would have been a significant development—if it had really happened. In fact, other medieval historians have rejected her interpretation, denying that any such relocation took place.[39] She misunderstood the fulling mill's impact, perhaps because she failed to observe that fulling mills were far less profitable than corn mills. That can be identified as the crucial factor determining where fulling mills were built, for only in districts where water resources were surplus to the requirements for milling flour did investment in a fulling mill make economic sense.[40]

William Langland, searching in the 1370s for a popular metaphor to express how a newborn child could not serve heaven until it was baptised and confirmed, chose that of newly woven cloth, unfit to serve as clothing until it had undergone a series of finishing processes. The first was that it had to be 'fulled under fote, or in fullyng stokkes'. As evidence, the order of Langland's words, putting the traditional method before the fulling mill, is too flimsy to trust; it is clear enough, though, that two centuries after the fulling mill's introduction he saw it as no more than an alternative to fulling by foot.[41] There is abundant evidence, too, that many people working in cloth production saw no advantage in the fulling mill. To take just one example: Salisbury during the fourteenth and fifteenth centuries was a prominent cloth manufacturing centre, well supplied with local waterpower. And whilst there were fulling mills at a little distance from the town, it is clear that much of the town's cloth continued to be fulled locally in the traditional manner: there were more than a hundred Salisbury fullers in 1421, evidently competing on grounds of quality and cost with the fulling mills.[42] Details from the ordinances of the fifteenth-century gilds of fullers found in the other major English provincial towns suggest that Salisbury's experience was typical, rather than unusual.[43] Such

---

[39] E. Miller, 'The Fortunes of the English Textile Industry during the Thirteenth Century', *Economic History Review* 2nd. ser. 18 (1965) 64-82; A. R. Bridbury, *Medieval English Clothmaking: An Economic Survey*, London, 1982, 16-25.

[40] Holt, *Mills of Medieval England*, pp. 155-8.

[41] W.W. Skeat, ed., *Langland's Vision of Piers Plowman: Text B*, Old English Text Society, O.S. 38, London, 1869, Pass. XV, ll. 444-50, p. 279.

[42] Bridbury, *Medieval English Clothmaking*, pp. 79-81.

[43] See evidence for continued use of traditional fulling methods in such major towns as, for instance, London, Bristol, York, and Lincoln: H.T. Riley, ed., *Memorials of London*, London, 1868, 400, 529; F.B. Bickley, ed, *Little Red Book of Bristol*, v. 2, Bristol, 1900, 10-16;

continued use of labour-intensive methods of fulling cannot be attributed simply to conservatism and to entrenched privilege. Traditionally, when historians have observed what seems to be the perpetuation of outmoded working practices, they have blamed the restrictive effect of craft gilds. But that is to misunderstand the distribution of economic power in medieval towns; master craftsmen, with or without a formal gild, were effectively the employees of merchant wholesalers, and would have found genuinely restrictive practices difficult to maintain. If fulling continued to be done in the traditional way, in a cloth industry dominated by powerful and shrewd merchants operating in an international market, then what has been presented as the superiority of the fulling mill is called into question.

It is to be regretted that writers on medieval technology have paid so much attention to Carus-Wilson's paper on the fulling mill, and ignored her other work on cloth production. In her survey of developments in the English cloth industry in the twelfth and thirteenth centuries, she identified a trend towards increasing organization of production by the dyers, who were extending their control over the rest of the different groups of workers in cloth.[44] Such a tendency was entirely consistent with what we know of other areas of enterprise in the Middle Ages: that the industrial craftsmen were generally only semi-independent of the merchants who marketed their wares; and that increases in production or perhaps just in profit were generally achieved through a reorganization of labour, of human resources, rather than through changes in the machinery or methods of production. This applied not only in the urban manufacturing sector, but also in the much larger agrarian sector. There were many improving landlords in the thirteenth century who looked for increased returns from investment in buildings, land improvement or changes in production; at the same time, however, the expedient of squeezing more from their tenants, in cash rents, labour services and miscellaneous seigneurial dues, was more generally resorted to, and seems to have delivered better results.[45] The reality of medieval production was that

---

Heather Swanson, *Medieval Artisans: An Urban Class in Later Medieval England*, Oxford, 1989, 40-42; Toulmin Smith, ed, *English Gilds*, Old English Text Society, O.S. 40, London, 1870, 179-82.

[44] E.M. Carus-Wilson, 'The English Cloth Industry in the Late Twelfth and Early Thirteenth Centuries', *Economic History Review* 14 (1944) 32-50, repr. in Carus-Wilson, *Medieval Merchant Venturers*, pp. 211-38.

[45] Edward Miller and John Hatcher, *Medieval England: Rural Society and Economic Change 1086-1348*, London, 1978, 233-9. For a study of the increased seigneurial demands being made on the peasantry, see Zvi Razi, 'The Abbots of Halesowen and their Tenants', in T.H.

most agriculture and virtually all industry was carried on in small units based on the household.[46] Under those circumstances, the introduction into existing productive units of new methods of working—even if any such were known—was a daunting task.

Exceptions were rare. Apart from the large-scale agriculture of the demesne, practised by English lords of the thirteenth and fourteenth centuries, though by few others, one example of large-scale enterprise was the Cistercian monasteries. Out of their concern to perform their own tasks, rather than live on rents or on the work of others, the Cistercians developed both large agricultural units, the granges, and industrial units on sufficient scale to make use of water-driven machinery. These Cistercian industrial mills are frequently cited as examples of medieval ingenuity, which indeed they were.[47] But they were exceptional, irrelevant to the mainstream of medieval production, just as the closed, celibate, privileged communities of the Cistercians bore no relationship of form or function to the basic social unit, the secular household. When the twelfth-century monks of Bordesley Abbey, for instance, chose to manufacture the assorted ironware they required with the aid of waterpower[48], it was because their overriding intention—driven by their strict interpretation of the monastic ideal—was to provide for their own needs, to contain production within their own precincts, and to achieve all this without hired labour. The cost effectiveness of the operation would not have been a consideration. What is most significant about this case is not that it happened at all; the technology of waterpowered hammers and grindstones was, after all, simple and easy to contrive. What we ought to note is that none of the other ironworkers of the region judged the technology to be appropriate to their own use. Likewise, urban merchants, the wholesalers of the principal products of the towns, did not follow the Cistercian model and form large units of production. They did not join together to form joint-stock companies, and build factories filled with their own machines, worked by their own employees. They were not yet attracted to industrial capitalism. On the contrary, in England just as elsewhere in western Europe, the merchant

Aston, P.R. Coss, Christopher Dyer and Joan Thirsk, eds., *Social Relations and Ideas: Essays in Honour of R. H. Hilton*, Cambridge, 1983, 151-67.

[46] Swanson, *Medieval Artisans*; E.M. Veale, 'Craftsmen and the Economy of London in the Fourteenth Century', in Richard Holt and Gervase Rosser, eds., *The English Medieval Town 1200-1540*, London, 1990, 120-40.

[47] Gimpel, *The Medieval Machine*, pp. 16-20, 56-60, 75-6; Reynolds, *Stronger Than a Hundred Men*, pp. 110-11.

[48] Holt, *The Mills of Medieval England*, p. 151.

wholesalers left production in household craft workshops where powered machinery had no place. It was the organisation and control of the mass of theoretically independent small-scale producers that they aimed for, and they achieved it through a variety of mechanisms: through the debt relationship, through borough law and custom, and in larger towns through the structure of merchant and craft gilds.

Improvement in productive techniques, and in the tools and machinery of production, did occur; and changes within the craft workshop inevitably made a greater impact on production than did the application of waterpower to single processes. The partial mechanization of fulling, for example, applied only to a single—and not the most time consuming—process within cloth production; the impact of the fulling mill was insignificant by comparison with that of the spinning wheel and the horizontal treadle loom which increased appreciably the work rate of the great numbers of men and women employed in the two principal processes of cloth production. Yet although the effect of both machines on the productivity of the individual worker must have been considerable, neither the extent of this change nor its social implications have been investigated. The period of the replacement of the warp-weighted loom and the vertical loom by the much faster horizontal loom is uncertain, and the process was probably a protracted one; but the new loom was in use in England and Europe by the late thirteenth century, and possibly well before that time.[49] The spinning wheel was an innovation of the thirteenth century, with both archaeological and historical evidence pointing to the fourteenth century as the period when it overcame spinning with the weighted spindle.[50] Unquestionably, the loom and spinning wheel reduced the costs of production in a way that the fulling mill could not have done. But not without cost to the workers: even a spinning wheel was expensive to purchase, and for many there must have been no alternative but to rent their machine or buy it on credit from the merchant clothiers, so increasing still further their indebtedness and their dependence on the men who organized their craft.[51]

Collections of essays on the range of medieval industries, drawing both on historical evidence and on the now substantial body of archaeological

---

[49] Derek Keene, 'The Textile Industry', in Martin Biddle, *Object and Economy in Medieval Winchester*, Winchester Studies v.7 ii, Oxford, 1990, 203-8.

[50] Bertrand Gille, 'The Medieval Age of the West', in Maurice Daumas, ed., *A History of Technology and Invention*, v. 1, London, 1969, 508; Swanson, *Medieval Artisans*, pp. 31-2; R.H. Britnell, *Growth and Decline in Colchester 1300-1525*, Cambridge, 1986, 75, 102; P.J.P. Goldberg, *Women, Work and Life-Cycle in a Medieval Economy*, Oxford, 1992, 145.

[51] Britnell, *Growth and Decline in Colchester*, p. 102.

evidence, have been edited by David Crossley and by Blair and Ramsay.[52] Together providing a variety of surveys of the major crafts, these separate studies demonstrate how insignificant a contribution powered machinery made to industrial production. Every analysis of the preponderance of the separate trades within a medieval town shows a consistent pattern: that broadly categorizing the principal occupations, most urban workers made a living from making and selling food and drink; from the primary and secondary processes of making clothing and footwear using wool and leather; and from the manufacture of metal goods. That pattern is true of large and small towns, in England as well as Europe.[53] And where did powered machinery contribute to those processes? In grinding the raw materials of the baker and brewer; perhaps in fulling the cloth after it was woven. It is also possible, though not very likely, that some of the tanners' bark was ground up in a mill, that the smiths' wrought iron had been smelted with the aid of waterpower, and that his finished blades were sharpened by waterpower. Sharpening mills are recorded, unusually, in a major English town, Winchester, where there were two in the fifteenth century; yet we are told that the streets of the city were littered with the cutlers' grindstones, most people still preferring to have their sharpening done more carefully by hand.[54]

When did powered machinery begin to make more than a restricted contribution to production? As a generalisation, it was at the very end of the Middle Ages, at the end of the fifteenth century and during the sixteenth century, that waterpower came to be usefully applied to a rapidly widening range of industrial processes in many parts of Europe.[55] This was the time when the iron industry of the Weald, the Black Country and elsewhere came to benefit from the new water-driven blast furnace: the first English blast furnace was built in 1496, and the technology spread all over southern and

---

[52] D.W. Crossley, ed., *Medieval Industry*, Council for British Archaeology, Research Report 40, London, 1981; J. Blair and N. Ramsay, eds., *English Medieval Industries*, London, 1991.

[53] See, for instance, Richard Holt, 'Gloucester in the Century after the Black Death', in Holt and Rosser, *The English Medieval Town*, p. 147; R.H. Hilton, 'Lords, Burgesses and Hucksters', *Past and Present* 97 (1982) 3-15, and repr. in Rodney Hilton, *Class Conflict and the Crisis of Feudalism*, revised edition, London, 1990, 121-31; Caroline Barron, 'The Fourteenth-Century Poll Tax Returns for Worcester', *Midland History* 14 (1989) 1-29; occupational analysis of 1,217 Nuremberg master craftsmen in 1363, in Friedrich Klemm, *A History of Western Technology*, London, 1959, 95-7, citing Hegel, ed., *Chroniken der deutschen Stadte*, v. 2, Leipzig, 1864.

[54] Holt, *Mills of Medieval England*, p. 152; Derek Keene, *Survey of Medieval Winchester*, v. 1, Oxford, 1985, 279.

[55] David Crossley, *Post-Medieval Archaeology in Britain*, Leicester, 1990, 139, and passim.

midland England during the century that followed. It was also in the sixteenth century that the rapidly growing edged tool industry of the Birmingham district was beginning to use waterpower for grinding and sharpening.[56] J.U. Nef long ago pointed to the new, large-scale enterprises which were appearing between 1540 and 1640, and it is noteworthy that all were capital intensive and often employed waterpower in one form or another. Manufactories of glass, paper, gunpowder, and the numerous mining enterprises of the sixteenth century employed dramatically improved and doubtless very expensive technology.[57] But although the number of such applications of waterpower increased with time, the historian needs to show caution in not exaggerating the extent of mechanization. It is impossible to evade the fundamental truth that no more than a minority of manufacturing processes were mechanized, even in part, before the nineteenth century. Historians must be cautious, too, in tracing the origins of these new techniques and processes. It is tempting, but surely mistaken, to assume that they must have evolved out of earlier enterprise or established practice. Not only does that project a misleading perception of the conditions of medieval or feudal production; it also serves to obscure the significance for productive technology of the far reaching economic and social changes of the early modern period. In the Middle Ages investment in mills came almost entirely from aristocratic landowners looking to increase their rents, a mentality quite unlike that of later industrial entrepreneurs. The expanding scale of post-medieval industry is a sign, too, of widening markets and real incentives to greater production.

CONCLUSION

In conclusion, we can observe that the medieval watermill and windmill saved a great deal of labour. All of the flour to make bread, and all of the malt for ale, had to be ground somehow—by muscle-power in the last resort. To what extent that would have been the muscles of horses rather than of people we cannot be sure. We also know that the price charged for grinding was regarded as a heavy one, even excessive; we are not certain, therefore, how far the ordinary people of the Middle Ages valued these powerful mills, or how much

---

[56] H. Cleere and D. Crossley, *The Iron Industry of the Weald*, Leicester, 1985, 106-8, 309-67; Richard Holt, *The Early History of the Town of Birmingham 1166-1600*, Dugdale Society Occasional Paper 30, Oxford, 1985.

[57] J.U. Nef, 'The Progress of Technology and the Growth of Large-Scale Industry in Great Britain, 1540-1640', *Economic History Review* 5 (1934) 3-24, repr. in Carus-Wilson, ed., *Essays in Economic History*, v. 1, 88-107; William Rees, *Industry before the Industrial Revolution*, 2 v., Cardiff, 1968, v. 1, 115-6, 121, 149-53, 181, and passim.

they resented them. Certainly there is no sign of the growing enthusiasm for the various forms of the mill which Lynn White and others have imagined. As yet, it is not possible to quantify the extent of the fifteenth-century move away from the watermill and the windmill in favour of the horsemill, although such a move was consistent with the antipathy of owners as well as of customers towards the expensive manorial mill.

We can also observe that there is no compelling evidence that medieval people made significantly greater use of powered industrial machinery as the centuries went by. Western experience during the modern period has been of social and economic development, associated with increased use of ever more sophisticated machinery in production, but that experience cannot be projected onto other peoples of other eras. Historians are still divided as to what precipitated the profound social changes of the later Middle Ages, although developments in technology can be ruled out as a prime mover of any significance.[58] The factors determining the number of cornmills were the size of the population and the degree of aristocratic coercion, a situation that still held true in the fifteenth century as it had done ever since the invention of the windmill nullified the availability of waterpower as a third factor. Nor can the number of industrial mills be easily related to expansions or contractions in the economy. Waterpower was an irrelevance to household-based, urban, craft production, especially in an era of cheap human labour and expensive raw materials. Limited markets and low volume production were the final factors ensuring that—with the exception of the fulling mill—minimal use was actually made of most of the small range of powered machines available to medieval people. The fact that mechanization at last began to be a serious proposition in a modest range of industries only during the later fifteenth century, as the medieval world was experiencing fundamental social change, simply serves to emphasize its limited role within the economy of the Middle Ages.

---

[58] T.H. Aston and C.H.E. Philpin, eds., *The Brenner Debate: Agrarian Class Structure and Economic Development in Pre-Industrial Europe*, Cambridge, 1985.

8

Agricultural Progress and Agricultural Technology in Medieval Germany: An Alternative Model*

Michael Toch

Anyone addressing medieval European agriculture and technology will do well to re-read chapter two of Lynn White Jr.'s *Medieval Technology and Social Change* of 1962. Despite distinguished forerunners like A. Meitzen (1895), Marc Bloch (1931) and George Duby (in an early article of 1954),[1] it was Lynn White who presented a theory lucid enough to become part of our understanding of medieval history and sophisticated enough to explain a very complicated process spanning at least three centuries. According to White, it was a technological revolution of the sixth to eighth centuries that broke the deadlock of low productivity and low living standards, thus making possible that dynamism and expansion we associate with the flowering of the Middle Ages.[2] Its main elements were: 1) horse traction instead of the oxen, with the concomitant innovations in harnessing, horse collar, horse shoeing, and wagon construction; 2) the heavy northern plough superseding the light Mediterranean one, which made possible the opening and draining of the heavy and water-logged alluvial European plains; 3) a new system of threefold rotation and new crops that both rationalised agricultural work and enriched human diet, as well as adding oats, the essential fodder-crop of the horse. Organisational changes completed this revolution, primarily the open field-

---

* I am much endebted to Professor Karl Brunner of the University of Vienna, who read a draft of this paper and let me share his intimate knowledge of Austrian social an economic history.

[1] A. Meitzen, *Siedlung und Agrarwesen der Westgermanen und Ostgermanen, der Kelten, Römer, Finnen und Slaven*, Berlin, 1895; M. Bloch, *Les caractères originaux de l'histoire rurale française*, Oslo, 1931/Paris, 1988; G. Duby, 'La révolution agricole médiévale', *Revue de géographie de Lyon* 29 (1954). Duby's *L'économie rurale et la vie des campagnes dans l'occident médiéval*, Paris, 1962, contained somewhat similar views to the ones put forward in the same year by Lynn White. It was translated into English, however, only in 1968 as *Rural Economy and Country Life in the Medieval West* (London, 1968).

[2] I am concerned here solely with the agricultural aspects of White's thesis, not with his 'stirrup' thesis regarding the feudal military revolution of the early Middle Ages.

system, which made possible a pooling of the resources necessary for such cultivation.

We shall not attempt to tackle this suggestive theory as a whole, although some of its elements must be questioned.[3] For the moment one should just point out that Lynn White Jr. commended what some of us today, more than thirty years later, might find difficult to applaud: for instance, the fact that the agricultural revolution of the early Middle Ages led to an estrangement between man and nature, a separation in which man became the master.[4] Such celebration, somewhat ambivalent and restrained in his book of 1962, is outright in a lecture delivered in 1963 and printed in 1967: 'We, who are descended from the peasants who first built such plows, inherit from them that aggressive attitude toward nature which is an essential element in modern culture. We feel so free to use nature for our purposes because we feel abstracted from nature and its processes'.[5] This tendency was even seen to possess a spiritual value, marking 'a major stage in our effort to master both the impulses within us and the forces and resources external to us'.[6] One might also mention that for White the feudal manor was but a co-operative association of peasants pooling their efforts for the common good: 'a powerful village council of peasants... whose arrangements were the essence of the manorial economy in northern Europe'.[7] Thanks to the agricultural revolution, peasants were able to live in 'big villages, with a tavern, a fine big church, maybe a school ..., certainly more suitors for your daughters, and not merely peddlers but merchants with wagons and news of distant parts'.[8] Power relations do not seem to enter into this picture at all. For all he took from Marc Bloch, White had no place for the medieval lord with all his formidable powers of coercion. If such a lord existed at all, he appears to have been an early prototype of the enlightened management of modern industrial corporations.

The early 1960s, when White wrote *Medieval Technology and Social Change*, were a time of strong belief in large-scale technology as the answer to

---

[3] Eventually I hope to present a much broader view of the questions raised in this paper in a volume to be titled, somewhat pretentiously, *A World History of Agriculture*.

[4] White, *Medieval Technology*, p. 57.

[5] Lynn White Jr., 'The Life of the Silent Majority', in *Life and Thought in the Early Middle Ages*, ed. R.S. Hoyt, Minneapolis, 1967, 85-100, reprinted in Lynn White Jr., *Medieval Religion and Technology, Collected Essays*, Berkeley, 1978, 133-47, here p. 145.

[6] White, 'The Life of the Silent Majority', p. 147.

[7] White, *Medieval Technology*, p. 44. This view is even more conspicuous in White, 'The Life of the Silent Majority', pp. 134-36.

[8] White, *Medieval Technology*, p. 67.

mankind's troubles, especially in North America.[9] These were also years of high employment, when industrial peace was at a premium and labour saving devices highly valued, as well as being the age of a 'new frontier' in space and elsewhere. Underlying this or resulting from it was a mind-set of almost unbounded optimism, a belief that matters are indeed manageable and developing from good to better.[10] By now, a generation later, technological development is soaring as never before, yet residual unemployment has become a fixed feature in all industrial societies. The outward thrust of Western industrial culture has effectively been challenged by non-Western societies as well as by mounting ecological and moral costs. As a result, our faith in technology—and its underlying cultural and political assumptions— has suffered much, forcing us in recent years to re-evaluate technology in a less optimistic way.

For that there is no better time than the central and later Middle Ages. European agriculture from the eleventh century to the early modern period is not noted for its technological prowess. To quote but one very influential author, Sir Michael Postan: 'the inertia of medieval agricultural technology is unmistakable'.[11] There were to be no inventions until the next agricultural revolution, roughly from the mid-eighteenth century onwards. So how did people cope with the problem of agricultural productivity during the twelfth and thirteenth centuries, when demand and prices for agricultural produce mounted, as well as during the later Middle Ages, when demographic adversity after the Black Death eased the Malthusian pressures and demand and prices dropped steeply?[12]

This study will contend that a) the picture of inertia is a misleading one, because b) the typical medieval form of agricultural progress was not

---

[9] For the essential North American quality and the unquestioning acceptance of White's very wording, one need only read the introductory pages of Norman F. Cantor, *Inventing the Middle Ages. The Lives, Works, and Ideas of the Great Medievalists of the Twentieth Century*, New York, 1991, esp. 22. Despite its sweeping title, this book is really about the introduction and impact of medieval studies in America.

[10] For a slanted and gossipy, yet revealing look at the academic antecedents of this mind-set, see Cantor, *Inventing the Middle Ages*, pp. 245-86. Cantor varyingly calls it 'progressivism', 'Wilsonianism' or 'rationality approach'.

[11] M.M. Postan, *The Medieval Economy and Society*, London, 1972, 44.

[12] The tacit use, for the sake of clarity, of yet another accepted view of medieval European economic development, namely the Postan thesis, is somewhat tongue-in-cheek. I believe the unified Neo-Malthusian view of late medieval demographic losses needs to be supplemented by a model that accounts for the wide regional variations. This view is to be presented in the introduction to my forthcoming *Die ältesten Rechnungsbücher des Klosters Scheyern (1339-1363)*, Munich, 1996 and in full in the study mentioned above in note 3.

technological change but rather the more intensive application of human work, just the opposite of the labour-saving devices and arrangements implicit in White's argument. A second main component was diffusion, organisational adaptation and elaboration, software, so to speak, rather than hardware.[13] Using some problematical pieces of White's theory, we shall first attempt to show that the question of diffusion was much more critical than the one of invention. We shall then deal with some concrete instances of agricultural progress and the lessons to be learned from them. We will draw our material mostly but not exclusively from medieval Germany. The basic idea, however, if correct, should be applicable to other parts of medieval Europe as well.

AGRICULTURAL TECHNOLOGY

A first problem concerns the timing and geographical range of White's agricultural revolution. Let us start with the horse, a major element in the theory. As in other European countries, in Germany the replacement of the ox by the plough-horse took not three but up to seven centuries.[14] Oxen were still normally used in early modern times, not just by backward peasants or die-hard conservatives, but also by the sixteenth and seventeenth centuries, agrarian reformers. For a very long time the horse served mainly in land transport. In agriculture its earliest employment was in harrowing rather than in ploughing. The discussion carried on in thirteenth-century English manorial management on the relative value of horse and oxen is exactly repeated in the agricultural treatises of sixteenth-century Germany.[15] The one's superior speed and traction force were again matched against the other's sturdy health and inexpensive feeding, not to speak of the oxen's superior value when slaughtered. In practice, different regions and different social layers within agricultural society used both animals, and only in the later Middle Ages are there (slight) indications for an increased employment of the horse. Not at all can we speak of the horse revolutionising arable cultivation in Germany at any point of time in the Middle Ages.[16]

---

[13] I owe this phrase to Karl Brunner of the University of Vienna.

[14] The latest and most important discussion on the topic is J. Langdon, *Horses, Oxen and Technological Innovation. The Use of Draught Animals in English Farming from 1066 to 1500*, Cambridge, 1986. See also W. Abel, *Geschichte der deutschen Landwirtschaft vom frühen Mittelalter bis zum 19. Jahrhundert*, 3rd ed., Stuttgart, 1978, 44.

[15] Ch. Parain, 'The Evolution of Agricultural Technique', in *The Cambridge Economic History of Europe*, ed. M.M. Postan, 2nd ed, Cambridge, 1966, 143-4 and Abel, *Geschichte* , p. 44.

[16] This accords with J. Langdon, *Horses*, p. 289: 'It would be very difficult to make a case that horse hauling was the sole factor behind this economic expansion'.

Another problem is posed by the scythe, a minor component by itself, yet one to which White gave great value for symbolising the fusion between 'Germanic animal husbandry' and 'Roman grain agriculture'.[17] There is much confusion as to the time and place of origin of this humble instrument of harvest. For White it was present already in Charlemagne's time.[18] Yet according to most other scholars, grains were harvested in the Middle Ages and beyond by the sickle, while the scythe was for a long time used only to cut the green fodder grass cultivated on meadows. According to French scholars, its adaptation to grain harvests came late in the Middle Ages, and only in 'progressive' countries where labour costs had to be considered.[19] An early evidence of such use can be found in Pieter Breughel the Elder's painting 'Corn Harvesting in Flanders' of ca. 1565 (Metropolitan Museum of Art, New York). The scythe's diffusion as a harvesting tool into Germany was to have taken place under the influence of such a progressive region as Flanders.[20] German writers, on the other hand, see the origin or at least the technical perfection of the hay-making scythe taking place not in the west but in the south-east, in the Alpine regions of animal husbandry and cultivated fodder meadows.[21] If the diffusion of important elements of the 'medieval agricultural revolution' went on right into the early modern period, we must be wary of accepting the time frame of Lynn White, namely between the sixth and the late eighth century. If there was an agricultural revolution, it should be placed later, in the period between the eleventh and twelfth century, as suggested by Georges Duby.[22] Only then could there have been a fit between the great reshaping of the German, indeed of the European landscape, by internal and external colonisation and, on the other hand, the spread of three-fold rotation, grain growing and open fields which White rightly associates as parts of one process. The manor as the main economic framework of agriculture and the lords' power over the peasantry should be added to the theory, for they too were part of the process. The lords were the only ones capable of enforcing on their property, demesne and peasants' holdings alike, the new division of land made necessary by three-fold rotation and open fields. They were the only

---

[17] L. White, Jr., in *The Fontana Economic History of Europe, vol. I: The Middle Ages*, ed. C.M. Cipolla, London, 1972, cited here according to the German translation, Stuttgart, 1978, 95.

[18] Ibid.

[19] Parain, *Evolution*, p. 156.

[20] Ibid.

[21] K. Hielscher, 'Fragen zu den Arbeitsgeräten der Bauern im Mittelalter', *Zeitschrift für Agrargeschichte und Agrarsoziologie* 17 (1969) 6-43, here p. 29.

[22] Duby, *Rural Economy*, pp. 90-99, 103-12.

ones who could direct the assarting and colonisation that brought new land under cultivation according to the new system. Finally, the lords must be part of the picture if only for their financing of the heavy equipment needed for this new type of agriculture.[23] The detailed research on agricultural implements and architecture carried out in Germany suggests at least two distinct cycles of technological diffusion.[24] The first lasted until the eleventh and twelfth centuries, and concerned mainly the basic technology of arable cultivation, such as deep ploughing and harrowing, harnessing (mainly of oxen and only marginally of the horse), and threshing (the flail). All of these were directed toward a more intensive utilisation of natural resources, soil, plants and animal. They had little to do with saving human toil. On the contrary, arable husbandry became increasingly labour-intensive, with extra workings (ploughing and harrowing) on sown fields as well as on the fallow. This made perfect sense in a time blessed by surplus population. Judging from the archaeological record, the agricultural implements of this earlier period were still mainly, often solely, made of wood.[25] There is yet little of the abundant iron which for Lynn White was such a clear sign of the early medieval agricultural revolution.[26] Labour-saving technology makes its appearance in German and indeed in European agriculture only in the later Middle Ages, when population was at a low. The use of the scythe for grain harvesting is an example, and a telling one in its implications. Unlike the slow handiwork of the versatile sickle, with which one can cut the plant's stalk anywhere, low down, halfway up or just below the ear of corn, the scythe cuts much wider and thus quicker, but low to the ground. Using a scythe meant giving up a certain portion of the harvest that could have been gathered by the more careful use of the sickle. This also meant that one had to forego a valuable by-product, the lower straw that would be eaten off by cattle, or taken away for thatch, litter or stall feeding, to name but a few of many uses. Thus increased speed was bought at a cost. Paying it was a viable option only when the

---

[23] M. Toch, 'Lords and Peasants: A Reappraisal of Medieval Economic Relationships', *Journal of European Economic History* 15 (1986) 163-82.
[24] Abel, *Geschichte*, 43-6; idem, 'Deutsche Agrarwirtschaft im Hochmittelalter', in *Handbuch der europäischen Wirtschafts- und Sozialgeschichte*, ed. H. Kellenbenz, v. 2, Stuttgart, 1980, 539-41; Hielscher, *Fragen*.
[25] Abel, *Geschichte*, p. 45; Hielscher, *Fragen*, pp. 13-14, 18, 20; as late as c. 1500 a Bavarian farm inventory still lists over 40 per cent of the tools made solely of wood, another 16 per cent of wood and iron, another 20 per cent of different materials, and only 16 per cent solely of iron: H. Sperber, 'Bäuerliche Geräte des Spätmittelalters', *Bäuerliche Sachkultur des Spätmittelalters*, Vienna, 1984, 291-306, here p. 294.
[26] White, *Medieval Technology*, pp. 40-41.

market could deliver those goods that had been previously manufactured as by-products of older processes of homestead agricultural production. This opportunity became widely possible only during the late Middle Ages.

Big spacious barns for threshing in inclement weather are another example of the rather few labour-saving devices that appear only towards the end of the Middle Ages.[27] Until then, threshing was done in the open and had to be finished early in autumn, thus requiring a large labour force. Appropriately, until the later Middle Ages there existed only smallish but strongly built granaries sufficient to shelter what needed to be sheltered: either the still unprocessed bundles of ears of corn cut by the sickle, or already threshed grains. Neither of them took up much room. Big barns to thresh under cover became necessary when harvesting with the scythe produced great quantities of sheaves that had to be protected against bad weather until the further processing that might go on right into the winter. They appeared only when the population shortage made gangs of hired labour employed at peak seasons too expensive. Larger barns in turn required a more substantial building technology and increased quantities of building material, for instance the flat straw used for roofing. But such straw had to be flailed, yet another work cycle, or bought. So one wonders whether these were really labour-saving technologies that made their appearance in the later Middle Ages. We should instead speak of yet additional adaptations, with rarely a new 'product' on the market. Even under the depressed demographic conditions of the late Middle Ages, such adaptations tended not so much to save labour as to reapportion its application in different ways.

AGRICULTURAL PROGRESS

Within the general gloom of a stagnant central and late medieval agriculture, scholarship has recently begun to identify pockets of agricultural progress. The first to innovate were the Dutch.[28] During the fourteenth and fifteenth centuries, they developed a system of cultivation of fodder crops, bed-and-row techniques for high value crops, heavy fertilisation and high labour inputs. High labour inputs mean, to spell it out, careful weeding by hand and spade cultivation, probably the most highly labour-intensive form of tillage ever in existence and a most ancient one. Such techniques were combined with the virtual elimination of the fallow and produced high yields per acre. None of

---

[27] Parain, *Evolution*, pp. 157-8; Hielscher, *Fragen*, pp. 35-8.
[28] B.H. Slicher Van Bath, 'The Rise of Intensive Husbandry in the Low Countries', in *Britain and the Netherlands*, vol. 1, ed. J.S. Bromley and E.H. Kossmann, The Hague, 1960, 130-53; H. van der Wee and E. van Cauwenberghe, *Productivity of Land and Agricultural Innovation in the Low Countries (1250-1800)*, Louvain, 1978.

them, except for the elimination of the fallow, were new. What was new was their combination and their geographical location near urban agglomerations.

Similar and even earlier clusters of progressive and highly productive agriculture were found in parts of thirteenth- and fourteenth-century Norfolk in England.[29] Here, progress took place in conventional arable husbandry rather than in the Dutch 'garden cultures', on manorial demesnes as well as on peasant holdings. Here too one key to increased productivity appears to have been the reduction of fallowing. Similar to the Dutch case, there was large-scale field cultivation of legumes. Their wholesome properties as nitrogen-fixing agents had indeed been part of agricultural know-how since Roman times. However, the high levels of productivity now attained make it very doubtful whether this knowledge had previously been as widely applied as White believed it to be.[30] Accompanying the expansion of legume cultivation was a reduction in pasturage and a heightened emphasis upon stall-feeding of livestock. Heavy fertilising made use of this supply of farmyard dung, as well as of other sources, such as marl, urban night-soil, and the droppings gathered in sheep-pens in neighbouring regions. It must have been these systematic methods of fertilising, together with the massive cultivation of legumes, that enabled this area to escape the vicious circle of declining soil fertility and falling yields elsewhere associated with English grain husbandry of the thirteenth and early fourteenth centuries. Another prime factor for the maintenance of agricultural productivity was the intensive use of labour. All the techniques mentioned and others like repeated ploughings and weedings meant substantial labour inputs, feasible until the Black Death of the mid-fourteenth century. Indeed, while the basic nature and scale of production remained much the same, the decline in productivity evident in this area in the fifteenth century can be directly attributed to a reduction in labour inputs. Norfolk was not the only English region to exhibit an unusually high degree of productivity. In other grain-producing regions similar techniques were applied: large-scale field cultivation of legumes, heavy fertilisation, heavy seeding rates, annual tillage.

For fourteenth- and fifteenth-century Germany, some redeeming features have been discovered, too. By then, agricultural production was not carried out anymore on the traditional large estate.[31] The lords had or were

---

[29] B.M. Campbell, 'Agricultural Progress in Medieval England: Some Evidence from Eastern Norfolk', *Economic History Review* 36 (1983) 26-46.

[30] White, *Medieval Technology*, p. 75: '... beginning in the late eighth century, legumes as field-crops came to play a vast and integral part in the new triennial rotation'.

[31] Duby, *Rural Economy*, 332-57; *Cambridge Economic History of Europe*, v. 1, Cambridge, 1966, 305-39, 581-632, 705-39; Kellenbenz ed, *Handbuch*, 542-44.

turning into rentier-landlords, their demesnes carved up and rented out to the peasants, and there came into being what has been termed the 'peasant economy' based on the family farm as the primary unit of production.[32] It was within this new economic and social environment rather than under the earlier tutelage of the manorial system that village life developed in its own right.[33] The German instances of agricultural progress can mostly be timed to the late medieval contraction in arable husbandry, the time of the 'agricultural crisis'.[34] They belong, in economic terms, to processes of conversion, whereby capital and labour are transferred from one sector in crisis (arable husbandry and grain production) to other branches of production for which there is a demand even under the general depressed conditions. Such new islands of productivity were located in so-called 'special cultures' (*Sonderkulturen*).[35] They were regional specialisations, first in wine growing, which experienced, in terms of acreage, the greatest expansion ever, at exactly the time when arable cultivation contracted.[36] Here as well as in horticulture, fruit-gardening, fishery, beer production, and the cultivation of industrial plants (flax, woad, madder, saffron), there was a quick development of expertise and techniques that cannot have been newly invented. Much of the experience in horticulture, fruit-gardening and pond-fishing had been known from the Roman authors copied in monastic scriptoria, but it was at this point in time that people began to put it into practice. Other branches were just as old, but here too existing practices were now applied in a more intense, although selective manner. The one really new feature, the development of a cotton industry in southern Germany, was no discovery at all but rather an application of foreign knowledge.[37] So here again there is diffusion rather than invention, the adaptation of previously well-known techniques to new circumstances. Labour and capital, not technology, were the variables in this game of combinations

---

[32] Th. Shanin, 'The Nature and Logic of the Peasant Economy' *Journal of Peasant Studies* 1 (1973) 63-80.

[33] For the political implications of this process see the work of P. Blickle, summarised in his 'Der Kommunalismus als Gestaltungsprinzip zwischen Mittelalter und Moderne', in *Studien zur geschichtlichen Bedeutung des deutschen Bauernstandes*, ed. P. Blickle. Stuttgart-N.Y., 1989, 69-84. For its cultural aspect, see M. Toch, 'Asking the Way and Telling the Law: Speech in Medieval Germany', *Journal of Interdisciplinary History* 16 (1986) 667-82. For its social implications see M. Toch, 'Emotions and Self-Interest: Rural Bavaria in the later Middle Ages' *Journal of Medieval History* 17 (1991) 135-47.

[34] For the concept in Germany see Abel, *Geschichte*, 110-49.

[35] For the following see Abel, *Geschichte*, 126-28.

[36] Exactly the same process taking place at the same time has been noted for French viniculture.

[37] W.v. Stromer, *Die Gründung der Baumwollindustrie in Mitteleuropa*, Stuttgart, 1978).

and permutations. In contrast to the manifold innovations in German commerce and industry, there is yet no research on the agricultural entrepreneurship of the later Middle Ages.[38] But there are good reasons to suppose that in Germany, as in the Netherlands, it was the pull of the urban market as well as urban entrepreneurship and money that provided the main incentive to the development of such specialized crop cultivation. This relationship has been proven, for instance, in the case of the sheep-rearing run along modernised lines near Augsburg, soon to become the town of the Fugger merchant princes.[39] Elsewhere it is the diffusion of sharecropping which allows a glimpse into the influx of urban capital into agriculture.[40] Not all the admittedly limited agricultural progress in late medieval Germany can be directly attributed to the modernising pull of towns. In animal husbandry and cheese production, a 'special culture' peculiar to the Alpine regions of Bavaria, Switzerland and Austria, intensification proceeded along secondary lines that are worthwhile considering.[41] Raising milk production is primarily a question of better feeding the animals. The most widespread methods for increasing food supply, noticeable since the late Middle Ages, were the following: the systematic gathering of wild hay in special containers along roads and in woods; a significant extension of summer grazing land in high altitudes (the so-called *Almen*), for which woods had to be cleared; the development of cultivated fodder meadows at high altitudes (*Bergmähder*); finally, no great surprise, the fertilisation of cultivated meadows at high altitudes (*Almanger*). Pure technological change is evidenced in one single case only, when in 1320 the ducal administration of Tyrol bought some 'big Hungarian cows' for breeding, without great success. Following this failed attempt, the systematic introduction of foreign cattle into the Tyrol had to wait until the seventeenth century. Besides regional intensive cultures, there were

---

[38] For the former see the extensive work of W. v. Stromer, most of which is listed in the footnotes to his 'Pionier-Innovationen und Innovationsschübe und ihr Einfluß auf Wirtschafts- und Lebensbereiche in Mittelalter und Frühneuzeit', in *Alltag und Fortschritt im Mittelalter*, Vienna, 1986, 121-30.

[39] R. Kießling, Bürgerlicher Besitz auf dem Land, in *Bayerisch-schwäbische Landesgeschichte an der Universität Augsburg*, ed. P. Fried, Sigmaringen, 1979, 135.

[40] For references and a discussion of this feature see Toch, *Lords and Peasants*.

[41] For the following see F. Treml, 'Die ostalpine Landwirtschaft vom 13. bis zum 17. Jahrhundert. Wirtschaftsformen und Erträge', in *Produttività e tecnologie nei secoli XII-XVII*, ed. S. Mariotti, Florence, 1981, 89-98. For the political and social implications see M. Toch, 'Voralpine Grundherrschaft und alpine Bauern im Spätmittelalter', in *Itinera* 5/6 (1986) 30-7; idem, 'Peasants of the Mountains, Peasants of the Valleys and Medieval State Building: The Case of the Alps', in *Montagnes, fleuves, forets dans l'histoire*, ed. J.-F. Bergier, St Katharinen, 1989, 65-70.

also local ones grouped in rings of gardens around the towns. Some of them specialised in one single culture, such as the cultivation of onions found all around Weimar, or of garlic around Nürnberg. All such islands of productivity within a sea of a stagnant agriculture exhibit a common denominator: no new technologies, but the intensive and systematic application of old techniques, sometimes combined one with the other, and all based on heavy inputs of labour. Such intensive application of human toil was also the rule during the later Middle Ages, when labour was expensive, and in all other branches of agriculture extension, not intensification, was the motto of the day.   In the next phase of demographic upswing, during the sixteenth and early seventeenth centuries, all the elements of progressive land management seen in the medieval English case were present in Germany[42]: heavy fertilisation, large-scale field cultivation of legumes and fodder-crops, a reduction of the fallow, although not to the degree practised in the Netherlands. But now there was also a new element, a certain willingness to experiment, as evidenced in the writings of educated administrators and pastors reflecting on the state of agriculture in their regions.[43] Some of them recommended to farmers new types of seed and breeds and more conscientious methods of cultivation. Others ran 'model farms' on which practices of intensification were put into use. One such farm belonged to Conrad Heresbach, a lawyer and ducal advisor who was also an agricultural author. He cultivated legumes, applied diverse fertilizers and experimented with combinations of crop rotations. Interestingly enough, he was one of the first to pay attention to the late Roman harvest machine mentioned by classical authors. His own clearly progressive practices, however, did not include new machinery. The only hint of hardware in the early modern period is a more frequent evidence of those iron plough-shares, iron-nailed harrows, and generally more handy utensils to which Lynn White had credited, prematurely it seems, the 'early medieval agricultural revolution' one millenium before. Even in the process leading up to the modern agricultural revolution, adaptations in crops, breeds and cultivation techniques were, initially, to be more important than new tools.

To sum up: medieval and even early modern technological change seems to have been much less revolutionary than it appeared to champions of technical determinism.[44] The main sequence of events needs to be revised to

---

[42] Abel, *Geschichte,* pp. 150-200.

[43] Abel, *Geschichte,* pp. 162-68, 190.

[44] This phrase is borrowed from the title of a review of White's book by R.H. Hilton and P.H. Sawyer, 'Technical Determinism: the Stirrup and the Plough', *Past & Present* 24 (1963) 90-100. I have my reservations about the validity of the two authors' highly critical censure, not least because of its undertone of insular defensiveness. For more sober criticism see J.Z.

accommodate for much longer processes of diffusion and adaptation, not to speak of regional variations. As the prime moving factor of productivity throughout the Middle Ages and parts of the early modern period, there appears the lavish application of human labour. This is no great surprise in times of demographic abundance. However, as the German and Dutch examples have shown, even in times of demographic crisis the heavy application of labour was possible if the market, usually an urban one, demanded so. The early medieval model of agricultural progress was supplanted by a second, late medieval model grounded in an institutional environment where peasants enjoyed a measure of freedom from the coercive power of lordship. At that time, the agrarian economy employed more rather than fewer people, relying on small-scale adaptations, transfers and conversions of resources, capital, skill and functions in systems small enough to be self-evaluating, self-controlling and self-regulating, yet tied to a larger environment by strong market forces and improved means of communication.

Is there something for us of the late twentieth century to learn from such history? Since the 'Postan thesis' on the exhaustion of medieval English soils, one should be aware of the fact that the application of a grain-producing medieval monoculture was definitely harmful to human beings as well as to the soil.[45] The Mediterranean-Arab tradition of oasis and garden cultivation, small scale agriculture, software agriculture if you wish, was largely destroyed in late medieval Spain by religious politics and a sheep-raising monoculture. Does it make sense today to import hardware, labour-saving Western technologies, into developing countries where there is unprecedented population growth and unemployment, destroying in the process what remains there of indigenous agricultural systems? Why must the penetration techniques of the heavy plough of medieval Europe and the American West be applied again to the last frontier left in the Amazon? Turning necessity into a virtue, might not the hungry billions of the world, might not the first world also be better served by that other small-scale model of medieval agricultural progress we have tried to sketch here? Adopting it might help to free ourselves from the relentless, aggressive and coercive attitude of modern man toward nature.

Titow, *English Rural Society 1200-1350*, London, 1969, 37-42, and Langdon, *Horses*, 288-90.
[45] Postan, *Medieval Economy and Society*.

Part III

Iron and Steel

9

Wood, Iron, and Water in the Othe Forest in the Late Middle Ages: New Findings and Perspectives

Patrice Beck, Philippe Braunstein, and Michel Philippe

(translated by Michael Wolfe)

To understand the true scope of the iron industry in medieval France, a society where existing documents often do no more than vouchsafe the role of artisans, it is necessary to trek through forests and along river banks where archaeological evidence of metallurgical activities can reveal both the potential and actual development of natural resources. Despite the shortcomings and inevitable distortions of written records before 1500, medievalists have dutifully sifted through the extant property ratifications and gifts of land and revenue rights belonging to ecclesiastical institutions, as well as cases of arbitration and final judgment of conflicts between competing claimants to water and wood. Among such sources, account books alone might permit us to grasp the actual operation and productivity of businesses that undoubtedly responded to complex market forces. Although the study of work sites allows us to hypothesize about their productive capacity, measuring the impact of mining and metallurgical industry on the medieval countryside requires that existing archaeological evidence be solidly dated before drawing any conclusions about true levels of production.

In approaching the phenomenon of industrial activity through the analysis of both texts and terrain, the Othe forest—located on the border between Champagne and Burgundy—offers special advantages. There is no lack of older published studies or on-site investigations; what has been lacking, however, in both the published and unpublished materials, has been a broader sense of perspective.[1] The first step toward rectifying this situation

---

[1] This article originally appeared as 'Le Bois, le fer et l'eau en forêt d'Othe à la fin du moyen âge: bilan et perspectives', in *Cahiers du Centre de Recherches Historiques* 9 (1992) 1-13. D. Cailleaux, 'Les religieux et le travail du fer en pays d'Othe', in *Moines et métallurgie dans la France médiévale*, P. Benoit et D. Cailleaux, eds., Paris, 1991, 193-211, presents a map detailing the essential topographical evidence culled from the work of T. Boutiot, A. Huré, and M. Quantin. Other seminal studies include A. Longnon, *Documents relatifs au comté de Champagne et Brie (1172-1361)*, v. 1, *Les fiefs*, Paris, 1877, v. 2, *Le domaine comtal*, Paris, 1904; A. Huré, 'Origine et formation du fer dans le Sénonais. Ses exploitations et ses

came as the result of a detailed analysis of a series of managerial account books for three large forges built in 1372 by the Countess of Flanders on her castellany in Champagne. These business accounts reveal the entire production cycle, from ore extraction to the final sale of market-ready iron. In the process, they illuminate the social relations between peasants, ironworkers and other lessees of comtal rights, administrators and iron merchants.[2]

As a point of comparison, our investigation focused particular attention on the managerial account books of other forges on the neighboring estates of the bishops of Troyes. The scope of the study was then widened to encompass the whole forest as well as other time periods during which mining took place. Initial findings from on-site surveys surprisingly revealed that mining activity in the forest stretched from the Iron Age to the very end of the Middle Ages.[3] Once the initial archival research and fieldwork had essentially been completed, a report was published that outlined our tentative findings and future directions of research.[4]

On-site studies concentrated primarily on a systematic investigation of wooded highlands and those fields and valley lowlands suggested by archival, cartographic, and oral evidence. More than a 1,000 hectares were carefully surveyed; 140 sites, most of them forested—sometimes heavily so—were plotted in and around Aix-en-Othe, Boeurs-en-Othe, Dilo, Rigny-le-Ferron, and Pâlis. Some sixty sites were discovered, more than half of them probably mines due to the frequent presence of iron ore and slag mounds found in the excavations and convex formations typical of the area's terrain. Of the more than 100 now registered slag heaps left over from mineral refining, some 75 still retain scoriae deposits, while there exist another dozen or so eroded mounds that are likely to be slag heaps.

---

fonderies dans l'Yonne', *Bulletin de la Société des Sciences historiques et naturelles de l'Yonne* 73 (1919) 33-106, which only considers the western portions of the forest; M. Quantin, *Cartulaire général de l'Yonne*, v. 1 and 2, Auxerre, 1854/1860, and *Recueil de pièces pour faire suite au cartulaire général de l'Yonne*, Auxerre, 1883; and A. Roserot, *Dictionnaire historique de la Champagne méridionale des origines à 1790*, Langres, 1945.

[2] P. Braunstein, 'Les forges champenoises de la comtesse de Flandre (1372-1404)', *Annales économies, sociétés, civilisations* (1987) 747-77.

[3] P. Beck, P. Braunstein, C. Dunikovski, and M. Philippe, 'La sidérugie ancienne en forêt d'Othe', in J.P. Metailié, ed., *Proto-industries et histoire des forêts*, Toulouse, 1992, 301-16.

[4] We wish to thank the Centre de Recherches Historiques for helping to make this study possible and the CIHAM in Lyons for local arrangements.

METALLURGY IN THE OTHE FOREST

Areas Identified with a high Density of
Mines, Slag-heaps and Forges

Figure 9.1. The Othe Forest in the Late Middle Ages

The weight of the slag mounds found at these sites, as determined by cubic measurement, generally hovered around several thousand tons.[5] The largest mound, overlooking the village of Bérulle, covered an area some 300 m in length, 20 to 30 m in width, with a height at its center of 3 m. It still probably retains some tons of iron slag. Although they paled in comparison, two other places also stand out with totals of 7,000 and 5,600 tons respectively and six others between 1,000 and 3,700 tons, with another three sites at 700 tons. The rest of the slag mounds amount to less than 500 tons. Some are very small, measuring no more than a few meters in length and retaining less than ten tons.

A number of iron sites appeared in remote locales, though this was probably due to the haphazard nature of prospecting, especially across the thickly wooded hills of the vast Rajeuses and Venizy forests that separate the clearings at Arces-Dilo and Boeurs-en-Othe. For the most part, however, the sites formed a loose chain along the forest edge in districts today crisscrossed by roads leading to tiny hamlets, farms, and abandoned watermills. Everywhere, place names in the area harken back to mining and metallurgical activities. For example, in the highlands above Aix-en-Othe, sites for mining and refining iron ore are dotted along the edge of the communal woods bordering the Brosses forest. One of these places is named La Forge, from which the Fourneau [Furnace] road, separating the 'fields of Fourneau' from the Forge de Fau, makes its way eventually to the hamlet of La Vove, located on the Nosle river, where an old watermill still stands today. Many of these same characteristic features of early iron industry can be found on other slopes of the valley where, again, readily available hydraulic power provided energy for all sorts of productive enterprises during the late Middle Ages and early modern era. In fact, where the woods of La Rachée meet the fields, in an area suggestively named Buisson Brûlé (The Burned Thicket), there was discovered a slag mound along with abundant chunks of iron-bearing sandstone.

Based on the considerable evidence provided by the slag mounds, some of them often located in very rough terrain, particular attention was given to cataloguing samples with a view to testing them more fully later on.

---

[5] Cubic volume is calculated using a formula based on spherical sections. Tonnage has been derived by using the methods proposed by G. Sperl in his *Uber die Typologie urzeitlicher, frühgeschichtlicher und mittelalterlicher Eisenhüttenschlaken*, Vienna, 1980. Our findings echo those formulated for the Morvan region by M. Mangin in his essay, 'Les mines et la métallurgie de fer en Gaule romaine: travaux et recherches', *Latomus* 47 (1988) 42, where he reports 'small iron mines of one hundred to one thousand cubic meters'.

This stage consisted of gathering samples of iron slag and, where possible, of ore. Overall, slag sites displayed amounts of heavy waste material laced throughout with iron, though occasionally from the heaps came lighter samples of scoria in a bubbled, spongy, and vitrified state. While direct processing characterized iron metallurgy in the Othe forest, as it did in neighboring lands, archival evidence suggests the early introduction of advanced methods of indirect iron processing. The presence of dark or vitrified dross in the upper level of a slag mine possibly reflects this technical change. Other kinds of evidence also help the historian reconstruct the lost industrial landscape. These include topographical clues or the presence of aquatic plant life suggestive of ancient hydraulic equipment; the faint traces of old roads; abandoned quarries originally dug out to meet local demand for flint, marl, and limestone; charcoal ovens; or the foundations of old shacks.

Establishing a correct chronology for the wealth of data found at these archaeological sites is crucial in order to date as precisely as possible the different stages of iron metallurgical activity extending over thousands of years. Only in this way can an actual place be properly identified in the written record. For this purpose, the investigation therefore had to focus on only those sites whose archaeological findings corresponded with archival information. Before reviewing what the written sources reveal about the history of late medieval iron metallurgy in Champagne, it is worth noting that our initial high hopes of combining these two approaches sprang from the archaeological study in the 1989 survey of a slag mound site believed to be one of those described by witnesses in a law suit between the abbot of Pontigny and the bishop of Troyes. Yet care must be exercised, since our search for mines in Pâlis following ironmasters' accounts from the late fourteenth century, yielded only sites from the Iron Age. One must therefore guard against the common urge simply to transpose onto a map those textual clues or bits of archaeological evidence that seem to suggest that a particular estate, given one or more references, necessarily discovered and then exploited local ore deposits. Our method tried to avoid these red herrings. The map found in this article merely presents in cartographical form the results of our regional survey, which loosely identifies the five main areas of iron production in the area, all of which can be cross-checked with written sources, though not in such a manner that the plentiful on-site evidence can yet be accurately dated as late medieval.

Working backwards may help solve some of these problems. It is commonly accepted that iron metallurgy in the Othe forest was just about defunct by the end of the fifteenth century, surviving in only a few select places into the first decades of the sixteenth century. Ironworks lasted the longest along the banks of the Vanne river; indeed, some of the original buildings that later served other businesses can still be seen today. The standard argument advanced to explain this decline, namely the exhaustion of local mines, only makes sense when placed in its broader context. Among the factors that must be accounted for were the cutthroat competition between firms using new methods of iron metallurgy and the rapidly disappearing ones still wedded to traditional processes. Also important was the fact that rising demand for lumber and firewood brought in more revenue from forest resources than the forges. All of these conditions conspired to cut short the development of even the most innovative ironworks. Although iron metallurgy in the Othe forest, subject as it was to European-wide changes, could not adapt to these new conditions, it is still worthwhile to broach the question of innovation, even though written sources almost invariably dwell much less on the ending of such stories than they do on their beginning.

This caveat should be kept in mind when using archival materials to piece together the history of iron metallurgy in the Othe forest during the late Middle Ages. The gradual shift in the fourteenth century of ironworks away from the wooded highlands to the waterways formed by the Vanne river and its tributaries, the Ancre, the Nosle, and Sévy creek, can be measured by the growing number of forges and mills for sharpening iron implements alongside those geared to grind grain, strip bark, process hemp, or make paper. These various establishments were leased or rented out from huge princely or ecclesiastical estates held by the bishopric of Troyes, the Duke of Burgundy, and the Cistercians, all of whom relied heavily on strong patrimonial control of forests. Iron forges and mills also occasionally appear on the lands of local landlords and royal officers.

As should be expected, word usage in the documents can sometimes convey a confused sense of technological practices, since signs of innovation must often be inferred from the writings of administrators and notaries. The term 'forge', for example, is the most widely used but also the least precise, as it could describe both forest-based metallurgical enterprises (the Hube forge, the forge of Mont-Erard) and those along rivers. In this latter case, it could designate a blacksmith's forge, the forge built in the late fourteenth century by the Countess of Flanders, or the iron refinery in Aix from around 1450, which was but a subsidiary plant related to the indirect method of iron processing. By this time, a new term—'foundery' (*fondoire*)—emerges marking a clear shift toward new techniques of iron processing. It appears first at Dilo in 1456 in the phrase 'waterfall for a foundery' (*saut à faire fondoire*), then at Gerbeau in 1477, and finally in a description of a foundery dismantled at Craney in 1515-16. Also indicative of this change was the shift away from the term kiln (*four*), found for example at Maraye in the fourteenth century and at Craney in 1412, in favor of the word furnace (*fourneau*), already evident in Normandy by the late fifteenth century.[6] Awkward attempts to be precise in such descriptions perhaps reflected the novelty or rarity of these operations, as in, for instance, the mention of a 'foundery furnace' (*fourneau de fondoire*) or 'iron smelting furnace' (*fourneau à faire le fer*) on an estate in Aix-en-Othe in the early sixteenth century. These were productive workshops, judging from the hefty rents of twenty to twenty-five *livres tournois* that they commanded. Nevertheless, they soon fell into desuetude at Craney, Chennegy, and Gerbeau, where they eventually became converted into grain or paper mills.

Experiments with indirect methods of iron manufacture in the Othe forest formed part of a larger development that, over the fifteenth century, left its mark on other places in Champagne, such as Rimaucourt and the areas around Saint-Dizier and Joinville, and adjoining regions in Burgundy at Bèze and Renève; at Bley and the Romaine valley in Franche-Comté; and at Les Andressis in Puisaye. An attempt should be made to explore the origins of these innovative methods. The land register of the seigneurial estate of Aix-en-Othe, drawn up in the mid-fifteenth century, clearly indicates that the 'forge to smelt iron' (*forge à faire le fer*), also called the 'hammer and refinery' (*marteau et affinière*), built on the banks of the Nosle river above a mill for 'sharpening iron implements' (*esmoudre ferrements*), was in complete ruin and had been out of service for five to six years. No other explicit mention of this plant can be found in the episcopal account books, though admittedly there are gaps in this register for the mid-fifteenth century. But, further down the Vanne valley, the first recorded instance in the area of using indirect methods of iron processing occurs nearly eighty years earlier at the forge of Chicheré in Saint-Liébaut—a site contemporaneous with those in Berry and the Saône valley.

During the 1370s, Nicolas de Fontenay, lord of Saint-Liébaut, bailiff of Troyes, and chief ducal officer after Saladin d'Anglure, took over control of the lands of Chennegy and Valcon from Guy de Pontailler, marshal of Burgundy, who had himself been involved in metallurgical industry in the Saône valley. In 1377, Nicolas de Fontenay signed with the monks of Dilo a ten-year lease for their mill at Chicheré, where he agreed to build 'two mills, to wit, a fulling mill near Loigny and another one for stripping bark near Saint Liébaut'. Ten years later, the monks averred that none of this promised work had been done, and that their original mill stood in ruin. What had happened

---

[6] J.F. Belhoste, Y. Lecherbonnier, M. Arnoux, D. Arribet, B.G. Awty, and M. Rioult, *La métallurgie normande (XIIe-XVIIe siècles). La révolution du haut-fourneau*, Caen, 1991, 47.

was that, upon signing the agreement, the lessee had redirected the water course in order to construct a 'forge for refining iron', complete with 'water wheel' (*grant roue*), all of which was managed by a certain Collesson le Liégois. In 1378, Nicolas de Fontenay took out a contract on mineral rights to the lands of the lordship of Aix and, in 1380, acquired from Saladin d'Anglure a parcel at Chasoy, near Villemaur, a large pond at Valcon, a mill, and some 400 acres of woods. He thus assured himself of both the ore and the fuel necessary for his business.

Hesitant to confront such a powerful man, the monks prudently waited until Fontenay's death before seeking to redress their grievances. This came in the form of a royal writ ordering the destruction of the water wheel and the restoration of the site with two mills for fulling cloth and stripping bark. It also required Fontenay's heir to pay damages of 120 gold crowns. His youngest daughter eventually became owner of properties in Chennegy and Valcon that included the water forge at Saint-Liébaut, which became a local placename by 1429 when residents referred to it as the 'ford at the forges' (*gué des forges*). Yet by then the ironworks had been relocated, since a century later the lords of Saint-Liébaut still operated a complete iron processing plant in the Ancre valley. A forge at Valcon produced 250 iron bars as late as 1519, while in 1520 a 'waterfall and place in Chennegy where long ago an iron foundery existed' was put up for rent. A merchant family named Boucherat, originally from Troyes but now based in Paris, had business interests in Valcon and once referred to the 'iron furnace' at Aix. However, these sites represented the last remaining ironworks in Champagne west of Troyes. Only the plant at Vandeuvre-sur-Barse still supplied pig-iron for the regional market after 1520.

Outside the valleys, which so far have dominated this look at the collapse of a thousand-year old industry, five or six large groupings of mining and metallurgical activity appear on the map as can best be determined by extant records. Especially rich documents from the episcopal lordship in Aix reveal that deep within the wooded highlands there existed ironworks contemporaneous with the forges built by the Countess of Flanders. These sources permit us to analyse specific sites that, taken together, present a portrait of forest life during the late fourteenth century.

Among the properties belonging to the episcopal lordship of Aix is a place called Mont-Erard. This place name no longer exists, though evidence clearly situates it in the vicinity of the present-day hamlet of Mineroy. Topographical references in the documents make this clear, such as the mention of the 'forest-paths and breaks between the Nullon and Jarruyer wood that lead straight to Mont-Erard...' or the wooded outcroppings of Mont-Erard, which correspond today with the Cornées Laliat, Alexandre, and Cabourdin, that rise to a altitude of 232 m in the commune of Aix-en-Othe.[7]

Around 1370, a bustling work camp had sprung up at Mont-Erard, run by a 'director of forges and forests' based in the bishop's residence in Aix-en-Othe, which also served as the center for managerial oversight and a place to stockpile inventory. In this wooded property, the bishop cashed in on his right to thin the forest, selling the cut timber for firewood, construction lumber, vine-props, slats, and planks. Yet iron ore deposits also spurred an intense metallurgical production that soon brought in more revenue than the amount generated by timber sales. The iron was produced on site and kept in piles before being loaded into packsaddles for transit out of the woods, as witnessed in a 1370 property survey occasioned by the death of the bishop, Henri de Poitiers.

Henceforth, however, these lease arrangements were apparently abandoned, thus severing the brief development in tandem of wood and iron production. As a result, a new alignment of economic and social interests occurred in the forests that benefited cultivators and merchants from outside the district, people generally unfamiliar to the settlers, miners, and woodcutters who lived at the behest of the bishop along the forest's edge. Also appearing on the scene from time to time was a new group, the inhabitants of the Nosle valley, who claimed a right of commons to sell iron ore, 'if the bishop doesn't have it mined for himself or in his name'. They sold ore to entrepreneurs like the aforementioned Nicolas de Fontenay, who contracted for iron ore from Mont-Erard in 1378-79, or iron merchants from family businesses, such as the one run by the Chevrillon clan, who leased for payments in kind both forges and mills for grinding blades. The sale of woodlands often followed. Jehannin Chappelain, for example, purchased in 1379-80 timber rights to woods located in the outcroppings of Mont-Erard, while in 1382-83 Pierre Breton and Feliz Jehannel, who operated a forge at Surançon, bought fifty acres. A year later, Jehannel's heirs purchased another fifty acres. These incremental changes gradually worked their way into the valley below, although receipts for forestry products soon peaked in the years 1385-90, after which they steadily declined until virtually disappearing from the registers after 1430.

Ground investigation in an area evidencing a high concentration of metallurgical operations might confirm the time frame suggested by documentary references to mining activity. Failing that, the question remains

---

[7] Mont-Erard should not be identified with the place today known as Mont Saint-Benoit, but instead is located on the other side of the Nosle valley. See P. Braunstein, *Les forges*, p. 757.

of how to account for the revenues and transactions across an even larger area whose exchange networks are presently unknown. Two other examples, also drawn from the written record, illustrate these difficulties as well as the methodological advantages of future planned archaeological studies. One such place is the lands around Pâlis, located on the right bank of the Vanne river. This area supplied iron ore by cart not only to the forge at Villemaur, but also by 1370 to the forges at Maraye and Surançon. Half the property and judicial jurisdiction of Pâlis belonged in the late fourteenth century to the Broutières family. In 1371, a knight named Jacques de Broutières levied 40 *sous* in ore revenues—a sum equal to the money raised by his judicial prerogatives and a quarter of the total collected from the main land tax, the *tailles*. However, no mention of ore revenue can be found in two other audits of the fief, the first conducted in 1361 by Jehan de Dinteville for the Countess of Flanders, and the second submitted in 1393 by Jehan de Broutières to the Duke of Burgundy. Between these years, iron ore mined from the fief supplied the needs of these three large forges, although uncertainty remains about whether such mining would have occurred but for the specific needs of the lord's estate.[8]

Another example, again from the area around Mont-Erard, stems from a property dispute between the bishop of Troyes and the abbot of Pontigny. An inquest in 1539, copies of which can be found in the archives of both parties, recounts the details of a contest that had dragged on for three centuries, several settlements notwithstanding. By 1530, the dispute became so bad that the abbot of Pontigny had ash merchants who worked on his lands without permission thrown into prison and their goods seized. The documents generated by this interminable suit clearly show that the woods buzzed with all sorts of activity: providing ash on account to merchants in Sens and Paris, making charcoal, splitting wood, manufacturing wooden hoops, cask staves, and wine-props. One witness reported that the swineherds bought from his father supplies such as salt, oil, water, and iron. In the early sixteenth century, iron ore mining continued in the woods at Jarruyer, although it was no longer refined on-site but rather carted down into the valley to places like the Cosdon forge on the Vanne river. Nevertheless, as episcopal account books clearly indicate, the woods continued to be cleared, as settlers came in to work some

---

[8] The connection between natural resources and processing equipment is spelled out in a contract between the knight of Foissy and the monks of Vauluisant, which states that the monks will have access to the iron mines in the woods only so long as the furnace can operate. Archives départementales de l'Yonne, H. 724 (1198). See also M. Quantin, *Cartulaire général*, v. 2, p. 488.

415 acres, 200 hectares of which covered ground where once mining and smelting took place. Henceforth, the area lost its name 'Stove Valley' (*Vallée des Poëles*) and was instead called 'Great Valley' (*Grande Vallée*).

The main significance of the 1539 inquest lies in its delineation between the administrative jurisdictions of two lordships in an area where boundary disputes revolved primarily around control of natural resources such as iron ore. Second is what it tells us about local archaeology. Nothing that the witnesses testified about remotely suggests that any of them had ever seen the mines or forges actually in operation. One of them recalled that, seventeen years earlier, someone had laid down a boundary near the '[d]warf's iron pit, an ironwork with large pieces of iron bloom' (*ferry au Nain, ferrier de grans et gros monceaux d'écume de fer*). This particular site had long since closed down, as had all the others where only abandoned clay-lined iron cookers were to be found. Expressions like 'the place called the ironworks' is in this light significant. Since witnesses disputed among themselves the very names of this or that slag mound, clearly memory of the area's metallurgical heritage remained strong, even though it had been at least a generation since the ore cookers and forges in the forest, then known as the Vallée des Poëles, had operated. It might have even been longer ago had a preliminary survey of one area ironwork not yielded data indicating a date in the late Middle Ages.

Similar kinds of evidence arising from boundary disputes or contests over communal rights reveal much the same picture for other sites and time periods of metallurgical activity in the Othe forest. Take, for example, the judicial inquiry conducted in 1503-04 concerning the property and rights of the Abbey of Valuisant, situated along the roads dividing the Rajeuses woods from the woods belonging to the glassworks at Fournaudin, which describes the presence of 'plenty of mines'. Along the boundary between Cérilly and Rigny-le-Ferron, one could 'follow the property markers hammered in by the monks all the way... to the yoke-elm standing at the old ironworks near the road'. Compromise agreements, sometimes very detailed, between lords, such as the Sire de Brienne and Sire de Venizy, or among monks at monasteries like the ones at Dilo, Pontigny, and Vauluisant, often prove quite helpful in determining the topography of iron mining and processing in the area during the thirteenth and fourteenth centuries. Another lengthy document describing the roads between Mont-Erard and Berluvier no longer mentions slag but evokes the presence of operational ore cookers and forges.[9]

Having traced the points of contact between written and archaeological evidence for several sites of mining and metallurgical activity, our

---

[9] *Cartulaire de Pontigny*, folio 294, Bibliothèque Nationale, Ms. latin 5465.

investigation is now ready to move to the next stage. There is little chance of discovering new accounts revelatory of the Othe forest's capacity to produce iron, though some additional information from the late fourteenth century does exist. In areas where slag did not generate very much revenue, only a systematic survey of the remaining slag heaps will yield an estimate of overall iron production. It will also be necessary to date as exactly as possible how long particular ironworks remained operational. Only in this way will a more nuanced, site-specific understanding of local iron metallurgy be realized.

Archival sources as well as preliminary studies exist for several places that permit us to reconstruct the history of forest management in the Pâlis woods, episcopal forests, the woods of lay lords, and the monastic holdings around Séant (Bérulle), Boeurs, and Dilo.[10] Thorough study of these places has revealed the prominent place held by iron metallurgy. In trying to gauge how industrial activity and technical change caused declines and shifts in the local economy, the role of local peasants, lords, and merchants proved decisive. Understanding the political interplay between individuals and groups, the different facets of the industry, the factors militating against attempts at innovation, as well as day-to-day operations at the ironworks, demonstrate not only the key place held by the Othe forest in the regional economy, but also constitute an important chapter in the history of European iron metallurgy. It is hoped that this essay has furthered these larger aims.

---

[10] H. Rübner, for example, in his *Untersuchungen zur Forstverfassung des mittelalterlichen Frankreichs*, Wiesbaden, 1965, has mapped out the fallow lands of Villemaur for the year 1328.

10

Weapons of War and Late Medieval Cities: Technological Innovation and Tactical Changes

Bert Hall

INTRODUCTION

This paper argues a specific thesis: that south German cities were the places where hand-held firearms developed. Such weapons helped to transform the art of war in the transition from the Middle Ages to the Renaissance, but little attention has been paid to their origins. We shall argue here that the invention of small arms (a modern, but conceptually useful term) was something of an accident, the inadvertent outcome of efforts to develop firearms that would serve to defend city walls in case of attack. By about the 1460s a gun had been developed that was light enough to be fired by a single individual holding it at his chest or shoulder and at the same time powerful enough to be worth having in battle. There are various reasons why this line of arms development emerged: in technology there is an important, but poorly understood innovation in gunpowder making; but the politico-military background, where developments heightened the sense of threat that many cities seem to have felt, was equally important. Just as cities and their defensive needs provided the military and technological context for the invention of small arms, so too citizen-soldiers, urban militias, constituted the human context.

URBAN MILITIAS

Medieval cities, as a rule, raised bodies of armed men on whom they could call when they needed some form of military force. In the Iberian peninsula, such urban forces were highly trained and heavily armed, and they played a considerable role in the *Reconquista*.[1] In Italy, on the other hand, citizen militias came to be supplanted during the later thirteenth century by professional armies under mercenary commanders, as politically ambitious cities like Florence and Venice acquired, through commerce, the wealth

---

[1] See James F. Powers, *A Society Organized for War: The Iberian Municipal Militias in the Central Middle Ages. 1000-1284*, Berkeley, 1987.

necessary to hire armed force.[2] The fourteenth and early fifteenth centuries saw
the dominant influence in Italy of mercenary armies, soldiers whose
professionalism eclipsed the clumsy, amateur style of the citizen-soldiers from
an earlier period.[3] The political dangers of relying on hired soldiers so
impressed humanists like Bruni, Villani, and Machiavelli that they in turn
idealized the older militias and added a certain moral baggage to the question
of citizen soldiers *versus* professionals that has never been wholly shed.[4]

The forces that shaped militias and mercenaries south of the Alps had
little bearing in Germany, but internal stresses within the cities did shape the
nature of militia service. There the old Germanic obligation of all freemen to
bear arms was transformed, in cities, into the obligation of citizens to defend
the city. More often than not, however, this generic citizen's militia gave rise
to a militia based in the city's commercial guilds. The right to bear arms was
incumbent on all citizens, but the ability to use arms effectively was vested in
members of the guild militia, citizens whose political reliability was usually a
great deal higher than in the ordinary or citizen militia. The economic
distinction between citizen militia and guild militia made all the difference.
Guild militiamen were the more well-to-do citizens, better able to afford
superior military equipment and to train in its use. Supplemented by
mercenaries in times of war, the guild militias defended the city when it was
besieged, and in peacetime they acted somewhat like a modern civil police,
maintaining internal order on the streets.[5]

CITIES AND ARMS

A complete map of late medieval Europe's arms-producing regions would be
heavily dotted. Only some of the dots would coincide with urban clusters, and
even amongst urban centers that produced weapons, not all were
technologically innovative. This highlights a peculiarity of the arms trade.
Any aggressive ruler needed as much good-quality armament as he could get,
and this need prompted the creation of 'captive' supply industries and
restrictions on exports of arms. However, restrictive policies usually had the

[2] A far more nuanced picture is presented by Daniel Waley, 'The Army of the Florentine
Republic from the Twelfth to the Fourteenth Centuries', in Nicolai Rubenstein, ed.,
Florentine Studies: Politics and Society in Renaissance Florence, Evanston, IL, 1968, 70-
108.
[3] The classic study is Michael Mallett, *Mercenaries and Their Masters: Warfare in
Renaissance Italy*, London, 1974.
[4] For example, Machiavelli, *The Prince*, chs 12-15.
[5] E. Haverkamp-Begemann, *Rembrandt: The Nightwatch*, Princeton, 1982, 37-8. The
important study by Theo Reintges, *Ursprung und Wesen der spätmittelalterlichen
Schützengilden*, Bonn, 1963 was unavailable to me.

unwanted effect of sacrificing quality in favor of ample supplies. The 'captive' industries were rarely centers of technological innovation. Like wine, weapons can be produced anywhere that a minimal level of skills and materials can be assembled. But just as making fine wine involves more than merely squeezing grapes and fermenting them, so too fine grades of weapons come from rare skills and special materials.

It was difficult to achieve first-class standing without the competitive pressures of the international arms market—and the prices that such a market could command. Yet in a paradoxical twist familiar to economic historians from many other industrial settings, once one achieved the acme of success and was able to produce extremely fine weapons, there was a tendency to restrict further technological efforts. Why, after all, would the market leader want to innovate even further and risk his hard-won success? Consequently, genuine technological innovation, the sort that had the potential to alter established patterns of business and warfare, was often born in cities that sought entry into this market. Neither cosseted 'captive' producers nor eagerly sought-after market leaders, cities with secondary reputations for quality that were free to sell to any and all—these cities tended to pioneer new types of weapons.

Amongst the technological and market leaders, the urban centers of Northern Italy stand out, especially Milan, famous for its finely crafted armor. But Brescia and Bergamo, as well as Venice, Trieste, and Venice's maritime protectorate, Dubrovnik, all were involved in arms production as well, with Brescia and Venice seeming to stand out as technological innovators. North of the Alps, a southern German complex of armament cities is found in Bavaria and the Tyrol, with centers in Augsburg and Innsbruck—both centers of extremely fine armor—as well as Passau, Landshut, and Nuremberg. Nuremberg was not particularly noted for fine quality armaments, but it became the most important center for firearms development in the fifteenth century. In the Low Countries, there were in Flanders important centers like Bruges, Lille, and Tournai; in Brabant, there were Brussels, Malines, and Anvers; and elsewhere there were other cities, like Mons and Liège.[6] Liège, in particular, became a model center for the commercial production of high-quality artillery that was sold all over Europe.

CROSSBOWS AS PARADIGMS
Although cities could make arms for any market that existed, it is worthwhile to ask what sorts of arms city soldiers preferred to use at home. The generic

---

[6] Claude Gaier, 'Le Commerce des armes en Europe au XVe siècle', *Armi e Cultura nel Bresciano*, Brescia, 1981, 155-68.

answer is plain to see: urban militia soldiers preferred missile weapons, crossbows and longbows at first, then firearms. Guild militias in particular esteemed the crossbow above all other weapons. In 1384, for example, Amsterdam had only one guild militia, under the patronage of St George and using the crossbow as its main weapon and its symbol. By 1436, this had grown to two companies, plus a third that used the longbow instead of the crossbow.[7] As other cities developed guild militias, these too adopted the crossbow more and more. In many respects the crossbow was an ideal urban weapon. It was compact, but quite deadly at short range. Unlike the sword, the crossbowman could not be drawn into a fight where he might be bested by a more skillful opponent. As crossbows evolved, they required an increasingly sophisticated body of skills to create, and they came to be fairly costly. They did not, however, require immense amounts of skill or lengthy periods of training to operate. Unlike the longbow, where training began in early childhood and where continuous practice was a necessity even among trained users for minimal proficiency, an individual crossbowman could obtain levels of competence satisfactory for militia service with a few months of training.[8]

Maritime cities like Venice and Genoa came to use the crossbow in another way: as the chief weapon on board their vessels. Venetian maritime law of 1255 required that crossbows be on board all Venetian vessels in numbers and sizes depending on the vessel's tonnage. Throughout the next sixty years, regulations steadily increased the numbers of crossbows as well as the amount and type of defensive body armor required on various vessels. The law came to require on-board specialists able to maintain and repair crossbows. The legal ratio of crossbowmen to seamen continuously rose. By the late 1320s, Ramon Muntaner, a famous Catalan mercenary captain with a great deal of experience in naval and land combat could claim unequivocally that 'crossbowmen win battles'.[9] (Since Muntaner was a professional contractor supplying crossbowmen to the highest bidder, his statement may be taken as an endorsement, or even a form of advertising.) By the same era, companies of Genoese crossbowmen were at the peak of their fame, and the city exported their services. At the Battle of Crécy in 1346, mercenary

---

[7] Haverkamp-Begemann, *Nightwatch*, pp. 37-8.

[8] Thomas Esper, 'The Replacement of the Longbow by Firearms in the English Army', *Technology and Culture* 6 (1965) 382-93, see pp. 391-3 for training. Similar conclusions based on physical evidence from the sunken warship, *Mary Rose*, are put forth by Robert Hardy, *Longbow: A Social and Military History*, 3rd ed., Cambridge, 1992, 217 ff.

[9] Frederic C. Lane, 'The Crossbow in the Nautical Revolution of the Middle Ages', in his *Studies in Venetian Social and Economic History*, London, 1987, 161-5, quote p. 165.

Genoese crossbow companies in the pay of France opened the hostilities, albeit with unfortunate consequences for themselves and their employers.[10]

Although the image of crossbow infantry arrayed against heavy cavalry in open combat possessed a certain appeal (fig. 10.1), the effectiveness of crossbows in field warfare was always limited by their slow rates of fire. Cavalry's superior mobility gave it a constant edge under these circumstances. Knights naturally hated crossbowmen, and had originally sought relief through prohibitions of these infernal new weapons. The Second Lateran Council in 1139 banned all bowed weapons in warfare between Christians, and there was a nearly contemporary edict by Emperor Conrad II forbidding crossbows.[11] Needless to say, knights were obliged to seek more effective counter-measures in the form of stronger armor. Knights had once carried shields and wore coats of mail (*hauberks*) as a suitable defense against arrows or spears; during the thirteenth century, more and more plate armor begins to appear, and by about 1330 full plate armor, capable of protecting all parts of the knight's body, became the dominant form of battle armor for the wealthiest and best-bred combatants.[12]

Unlike the longbow, which also threatened knights, but which was effectively a fixed technology, crossbows continued to develop during this period, becoming more powerful and more threatening through a series of technical improvements. The simple wooden crossbows of the eleventh and twelfth century were displaced by more resilient composite bows made of horn, sinew, wood, and glue. These required stronger strings and more robust locking nuts and trigger mechanisms. Several mechanical methods of cocking the bow begin to appear, the increased stiffness of the main bow made spanning the weapon impossible without some assistance.[13]

Cities provided a superior environment, so to speak, for the crossbow as compared with open field warfare. In the latter, longbows always remained the preferred weapon, at least for those lucky enough to be able to muster such troops. In cities and in urban warfare—in the defense of walls, for example— the crossbow's drawbacks, such as heaviness and a slow rate of fire, counted

---

[10] Alfred H. Burne, *The Crécy War: A Military History of the Hundred Years War from 1337 to the Peace of Bretigny, 1360*, London, 1955, 169-92.

[11] Rosemary Ascherl, 'The Technology of Chivalry', in Howell Chickering and Thomas H. Seiler, eds., *The Study of Chivalry: Resources and Approaches*, Kalamazoo, MI, 1988, 263-311; see p. 276. J.F. Guilmartin, 'Technology of War', *Encyclopaedia Britannica*, 15th ed, revised, New York, 1991, v. 29, 529-47, see p. 539. (NB: This article is not in earlier redactions of the 15th edition; it appeared for the first time in the 1991 revision.)

[12] Kelly DeVries, *Medieval Military Technology*, Peterborough, 1992, 79.

[13] Jost Alm, *European Crossbows: A Survey*, H. Bartlett Wells and G.M. Wilson, trans, London, 1994, 19-41 and figs. 13-26.

for less than its advantages. Just as maritime cities found the crossbow ideal for shipboard use, so too many other cities found it well-suited for siege warfare. Here, of course, the association of the crossbow with local militia troops became even closer, for local militias made up part, often the bulk, of the fighting troops defending a city under siege. Wall defense work promoted the development of special heavy or 'rampart' crossbows, weapons so heavy as to require stands and even more extensive mechanical aids to span them.

Early in the fifteenth century, crossbows with steel bows make their appearance, achieving even greater penetrating potential in a smaller and more portable weapon. The steel crossbow was an expensive device, however, as tempered spring steel was a costly, skill-intensive material to make. Steel springs were the technical development that seems to have cemented the relationship between crossbows and cities, for those smiths able to make the annealed and tempered steels required for heavy crossbows usually preferred the freedom and income-earning potential available only in cities. Armorers responded with still stronger armor as they learned to exploit their metal-working skills to higher limits and make protective suits that were lighter and stronger all at once. Naturally, some cities—Milan and Augsburg spring to mind—became noted for the excellence of their armor, and the products of their smiths became the pride of many an élite wearer.

PICTORIAL EVIDENCE

The familiar dialectic of offense and defense meant that late medieval arms-making cities were involved in a continuous technological competition, simultaneously commercial and military, that defined their industrial as well as their political profiles. Little wonder then, that by the sixteenth century iconic images of weapons become familiar parts of a city's self-image and self-representation. Three examples from well-known sources will suffice.

Hans Sachs' poetic descriptions of the trades of the city of Nuremberg was published in 1568 as the *Ständebuch* (Book of Trades) with illustrations by Jost Aman.[14] Among the trades depicted is the armorer, able to make both public and 'secret' or hidden armor. (fig. 10.2) The *Pogner* or bowyer, makes crossbows. (fig 10.3) '*Gut Armbroster kan machen ich'*, says the legend, and further boasts of the bowyer's skill at making either horn bows or steel bows ('*Mit Huernen oder Sthaelen pogn.... '*). The Nuremberg of Hans Sachs contained a huge firearms-making industry, and the *Ständebuch* contains a *Buechsenschmidt*, a gunsmith, able to make long or short barrels for muskets

---

[14] Jost Amman and Hans Sachs, *The Book of Trades Ständebuch*, 1568, intro., Benjamin Rifkin, New York, 1973.

Fig. 10.1. The Fantasy: Crossbows in Open Battle

Der Panzermacher.

Fig.10.2. Panzermacher

Der Pogner.

Fig.10.3. Pogner

Der Büchsenschmidt.

Fig. 10.4. Büchsenschmidt

Der Büchsenschäffter.

Fig. 10.5. Buechsenschaeffter

Fig. 10.6. Popinjay Shooting

Fig. 10.7. Map of the Nightwatch

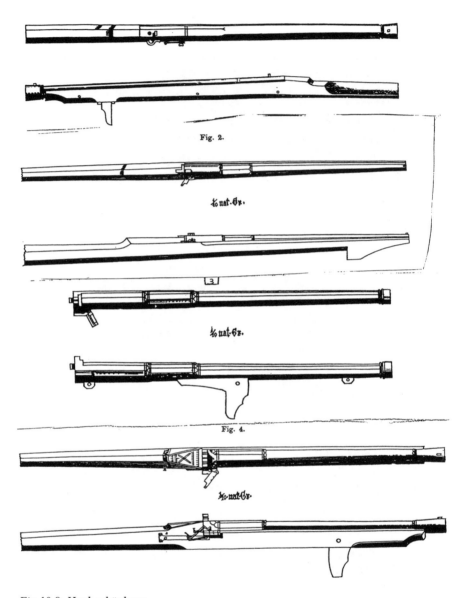

Fig. 2.

⅓ nat.Gr.

⅓ nat.Gr.

⅓ nat.Gr.

Fig. 4.

⅓ nat.Gr.

Fig.10.8. Hachenbüchsen

or pistols. (fig.10.4) (Note the pistols hanging above the forge.) Finally, in a nod to the sort of specialization that can only develop from in a mature industrial base, Jost Aman shows the woodworker who specializes in making gunstocks, whose poem tells us how he fastens the barrel to the stock. (fig. 10.5) 'Thus', the verse concludes, 'honest men can ward off robbers'. (Note here also the pistols on display above.) The *Ständebuch* clearly shows Nuremberg as a major center of arms production.

The relationship between militias and firearms is evident in a woodcut from Olaus Magnus showing the common form of target practice, parrot or popinjay shooting, aiming at a stuffed bird mounted on a pole. (fig. 10.6) Such public competitions were usually held around Pentecost, along with a banquet for the militia company and prizes for the best shots. It is interesting to note that the militiamen use a mixture of firearms and crossbows, which along with Jost Aman's illustrations suggests that there was a long period in which both weapons co-existed in essentially the same role. The public and ceremonial character of what amounts to target practice reminds us that skill with arms remained a defining characteristic of the individuals who made up the militia, and a symbol of the city's ability to maintain internal order and defend itself against outside threats.

The most famous representation of the militia's identification with missile weapons is Rembrandt's *The Nightwatch*. (fig. 10.7) In fact, the painting was not called by its modern name during Rembrandt's lifetime, but was named after what it represents, *The Militia Company of Captain Franz Banning Cocq*. Rembrandt accepted a commission to paint the Company in 1642 after the conventional manner of group portraits of militia companies.[15] (Perhaps a hundred or more such militia company portraits survive from the seventeenth century.) Records indicate that the sixteen militiamen who are depicted all paid commissions to Rembrandt that were based on the amount of their bodies to be represented and on their prominence in the finished painting. Capt. Banning Cocq and his lieutenant, Wilhelm van Ruytenburgh, get pride of place, followed by the Ensign, Jan Cornelius Vissher, and two sergeants, Rembout Kemp and Reijer Engelen. These people were all thoroughly respectable members of Amsterdam's ruling patriciate. Banning Cocq was already a wealthy patrician who had married into a titled family; he would eventually serve as an alderman and as Amsterdam's Burgomaster. Wilhelm van Ruytenburgh was almost as wealthy, and he also became an alderman in

---

[15] Haverkamp-Begemann, *Nightwatch*, pp. 9-20.

time, but never Burgomaster. The ensign and sergeants were likewise men of substance and some influence. Kemp, for example, was a deacon in the Reformed Church and noted for his charitable work among the city's poor.[16]

Rembrandt's is a group portrait of Dutch bourgeois respectability at its apogee, and yet *The Nightwatch* fairly bristles with guns. A map of the image as it was before it was trimmed down in the eighteenth century shows no fewer than six musketeers, five of whom survive today.[17] (fig. 10.7) The muskets do not have a central role in the ensemble because muskets were not carried by the high-status officers and sergeants who make up the social center of the portrait, but they are important nevertheless. The militia companies of Amsterdam were originally called 'shooting societies' (*schutterijen*), and their members were 'shooters' (*schutters*), while their meeting halls were 'target ranges' (*doelens*). Rembrandt's painting was originally meant for the *Klovenierdoelen*, the guild hall of one of the principle militia guilds in the city, going back to 1522. The building's name translates literally as 'the target range of the culveriners', and 'culverin' is one of many names in Dutch for a musket.[18] (The other chief militia guilds, by the way, were named for the longbow and the crossbow, although by Rembrandt's day they all carried muskets.)

The muskets symbolize the tradition of the guild to which Capt. Banning Cocq's company belonged, and they represented as well the sort of military strength that had helped the Dutch Republic defend itself successfully against Spain's efforts to dominate the Low Countries. The Amsterdam militia companies had been involved in military actions in 1622 (against Zwolle) and 1632 (against Nijmegen), events that were still living memories in 1642. When Amsterdam and the Prince of Orange came to the point of open conflict in 1650, the city closed and manned its gates against a planned siege headed by William II himself. Indeed, Capt. Banning Cocq himself reissued the rules for militiamen during the crisis. Guild militias became largely social institutions later in the seventeenth century, but they were still necessary and powerful groups at the time Rembrandt made his painting. The guns in the picture suggest how violent force was legitimated and harnessed for political ends in virtually all late medieval and Renaissance cities.

CITIES AND ARTILLLERY
The link between city militias and missile weapons serves as background for the more specific question of where and how portable firearms emerged. The

---

[16] Haverkamp-Begemann, *Nightwatch*, pp. 21-32.
[17] Haverkamp-Begemann, *Nightwatch*, fig. 2.
[18] Haverkamp-Begemann, *Nightwatch*, p. 42.

other necessary background feature concerns the ways that firearms changed how cities might maintain themselves as quasi-independent units in a shifting political-military environment. Cities depended on their ability to withstand sieges for a large part of their ability to survive. Plainly, the balance of force between defenders and attackers was of considerable moment to any city. It is hardly accidental that one of the very first documents to mention guns in Europe comes from Florence, where on 11 February 1326, the Signoria charged two magistrates with the tasks of 'making or causing to be made for this community iron shot for cannons', as well as bronze cannons themselves.[19] At first, however, firearms were not a serious threat. City walls were among the first targets of the new weapons, but cannons themselves were too few and too feeble to represent much more than a nuisance.

This situation was mainly a matter of raw materials, chiefly the difficult time Europeans had obtaining saltpeter, which does not occur naturally in most of Europe.[20] The saltpeter scarcity meant that gunpowder was expensive and only available in limited quantities. Scarce and expensive powder meant small guns firing weak shots. By the closing decades of the fourteenth century, this difficulty was solved. Europeans learned to cultivate saltpeter from decaying organic matter, and while the saltpeter they got was highly impure, it soon came to be cheap and plentiful. As it did, guns grew more numerous and also bigger in size. By the beginning of the fifteenth century Christine de Pizan argues for a roughly equal division of 'artillery' between trebuchets and guns in her redaction of Vegetius.[21] Henry V's campaign in northern France beginning in 1415 is marked by a new emphasis on firearms, guns meant to reduce stubborn cities to submission in short order. Sources comment on how heavily equipped with firearms Henry's expedition was when it sailed from Portsmouth in mid-August.[22] The French chroniclers

---

[19] '*Ad faciendeum et fieri faciendum pro ipso communi pilas seu palloctas ferreas et canones de metallo pro ipsis canonibus et palloctis habendis et operandis...*'. See Victor Gay, *Glossaire archéologique du moyen âge et de la Renaissance*, 2 vols, Paris, 1887, s.v. 'artillerie'.

[20] This discussion follows my article, 'The Corning of Gunpowder and the Use of Firearms in the Renaissance', in B. Buchanan, ed., *Gunpowder: The History of an International Technology,* Bath, 1996, pp. 87-120.

[21] Christine de Pizan, *The Book of Fayttes of Armes and of Chyualrye*, A.T.P. Byles, ed., London, 1937, 154-59. See Bert Hall, '"So Notable Ordynaunce": Christine de Pizan, Firearms and Siegecraft in a Time of Transition', in C. De Backer, ed., *Culturhistorisch Kaleidoskoop: Een Huldealbum aangeboden aan Prof. Dr. Willy L. Braekman*, Brussels, 1992, 219-40.

[22] Friedrich W.D. Brie, ed., *The Brut or The Chronicles of England*, London, 1906/08, v. 2, 382.

emphasize the 'unheard of size' (*inaudite grossitudinis*) of Henry's weapons, as well as their noise and smoke, during the siege of Harfleur.[23] Henry's success at Harfleur in 1415 (which led to the Battle of Agincourt) was the first in a series: Caen in 1417, Falaise later the same year, Cherbourg in 1418, Rouen in 1418-19, Dreux after less than a month of siege in 1421, and Meaux in 1421-22.[24] The powder expenditure required to reduce Normandy to obedience would have been considered ruinously expensive by the standards of an earlier age. Each firing of, let us say a 400 pounder, would have taken some 60 lbs of gunpowder, to say nothing of the powder consumed by the smaller firearms.

If the new, larger guns caused unease among city councillors, the south German cities of the Holy Roman Empire had their own special reasons to worry. The Bohemian heretics who followed Jan Hus after 1415 successfully defended their heterodoxy during the 1420s and 30s against the armed might of the Empire with innovative firearms tactics. The Hussite wagon fortress (*Wagenburg*), with its large complement of small cannons, was widely feared and equally widely imitated throughout the German-speaking world. After turning back several imperial armies, the Hussites made several military thrusts into neighboring Moravia and Saxony in the late 1420s. These failed to gain any new adherents to the Hussite cause, but they provoked a wave of panic and crusading zeal amongst the German cities. The city of Nuremberg expected to face being besieged by the Hussites, and in 1430 drew up a series of ordinances for the event. The Catholics determined to fight the Hussites in their own style, with an imperial *Wagenburg*, but this too came to grief at the Battle of Taus (14 August 1431).[25]

DEVELOPMENTS IN GUNPOWDER
The general threat artillery represented to cities, and the Hussite threats (both real and imaginary) to cities in the southern and eastern parts of Germany, make up the background to efforts to make smaller and more portable firearms. Small weapons could be used to defend city walls and in the

---

[23] L. Bellaguet, ed., *Chronique du religieux de Saint-Denys*, Paris, 1839-52, v. 5, 536-37.

[24] C.T. Allmand, 'Artillerie de l'armée anglaise et son organisation à l'époque de Jeanne d'Arc', in *Jeanne d'Arc: Une époque, un rayonnement*, Paris, 1982, 73-83, see p. 74. See also the classic study by Richard A. Newhall, *The English Conquest of Normandy: A Study in Fifteenth-Century Warfare*, New Haven, 1924.

[25] Volker Schmidtchen, 'Karrenbüchse und Wagenburge—Hussitische Innovationen zur Technik und Taktik des Kriegswesens im späten Mittelalter', in *Wirtschaft, Technik und Geschichte: Beiträge zur Erforschung der Kulturbeziehungen in Deutschland und Europa: Festschrift für Albrecht Timm*, Volker Schmidtchen and Eckhard Jäger, eds, Berlin, 1980, 83-109, see p. 85.

Wagenburg, turning the threat posed by gunpowder into an asset to be used against the attackers. But small, portable guns had always been pitifully weak, lacking either the range or, more important, the penetrating power, to do an adequate job. Technical changes also took place in the first decades of the fifteenth century that improved the performance of small firearms. Powder, as the name implies, was originally a finely-mealed mixture of nitrate, sulfur, and powdered charcoal. The impurities that were abundantly present in medieval nitrates made them highly prone to absorb atmospheric moisture, which meant that mealed gunpowders spoiled quite quickly. One way to extend the 'shelf life' of gunpowder was to wet the mixture and stir it into a thick paste. This could be shaped into round loaves (or 'dumplings' as one text calls them) that, when dried, kept the gunpowder inside the loaf shielded from the atmosphere. The loaves were crushed into powder again when the need arose. The earliest dated description of this process comes from 1411, but it was slow to be widely adopted. For a number of technical reasons this reconstituted, 'corned' gunpowder burned much more rapidly than the original mixture, and this made it problematic to use as a propellant in large artillery. Gunners were extremely cautious about adopting it for fear it would burst their guns, but they soon found that 'corned' powder was a superior form of gunpowder in smaller firearms. The next wave of technical development after the giant bombards that cheaper gunpowder made possible was towards smaller, portable firearms that corned gunpowder made practical.

THE 'HOOK GUN'
The main beneficiary of this technical trend was the German *Hackenbüchse* or 'hook gun'. This weapon was named for a projection on its under surface near the muzzle that allowed it to be steadied on battlements for firing. 'Hook guns' appear for the first time in 1418, shortly after the primitive descriptions of corning but before the Hussite revolt became a threat.[26] They spread dramatically throughout the German arms-making cities during the 1420s and 30s as corned gunpowder became more familiar and as the Hussites appeared more menacing. In their original form, most *Hackenbüchsen* were somewhat larger than a hand firearm; some even required a two man crew. (fig. 10.8) The tactical purpose of a *Hackenbüchse* was a firearm that could easily be aimed downward and moved from one porthole or arrow slot to another as the

---

[26] Johannes Karl Wilhelm Willers, *Die Nürnberger Handfeuerwaffe bis zur Mitte des 16. Jahrhunderts: Entwicklung, Herstellung, Absatz nach archivalischen Quellen*, Nürnberg, 1973, 4, citing Bernard Rathgen, *Das Geschütz im Mittelalter: Quellenkritische Untersuchungen*, Berlin, 1928; reprt. V. Schmidchen, ed., 'Klassiker der Technik', Berlin, 1987, 62.

situation might demand. It was meant to be the kind of weapon that could disrupt the besieger's gun crews and make it impossible for them to breach the city's walls. Because the attacker's guns were usually at this time protected only by light wooden screens or modest earthen mounds, the defender's guns could sacrifice sheer mass for portability and the possibility of a favorable angle to shoot from.

The *Hackenbüchse* quickly became a favorite weapon in city arsenals in southern and central Germany. By 1430, the city of Nuremberg conducted an inventory survey of weapons available for the city's defense. It showed some 501 hand guns, mainly *Hackenbüchsen*, as against 607 crossbows. At about the same time, a *Büchsenschützengesellschaft* was formed with the blessings of the *Stadtrat* to promote the production and use of firearms in Nuremberg. Other cities with less production capacity, like Frankfurt, purchased firearms from Nuremberg sources to meet their obligations under the special levies sent to combat the Hussites. By the mid-fifteenth century in Nuremberg, regulations for defense of the city's gates stipulated 20 firearms and 10 crossbows per gate. Another inventory of 1462 revealed 2,230 firearms ready for the city's defense; 2,052 of these were *Hackenbüchsen* or closely related types. By contrast, a comparable inventory carried out in Cologne in 1468 tallied only 348 firearms, and yet another inventory in Ghent in 1479 counted 486 (three-fourths of them hand-held).[27] Nuremberg was clearly the most heavily-armed of the quasi-independent German cities, and at the same time the center of an enormous arms-making industry that exported its wares to a number of customers. Nuremberg was a source of capital for mining interests in Eastern Europe, whose products became the raw materials for local skilled craftsmen, whose products Nuremberg then distributed throughout the rest of the Empire.[28] The city's large arsenal simply reflected its dominant position in the arms trade of the fifteenth century.

What made the popular *Hackenbüchse* a pivotal development in the history of arms is that it could be 'miniaturized', shrunk to proportions that made it suitable for a single soldier to use without hooks, struts, or other aids. By the middle of the fifteenth century, the matchlock mechanism had appeared, making it possible to ignite the gunpowder charge with only one

---

[27] Rathgen, *Geschütz*, p. 261 (Nuremberg) and p. 321 (Cologne). Claude Gaier, *L'Industrie et le commerce des armes dans les anciennes principautés belges du XIIIme siècle à la fin du XVme siècle*, Paris, 1973, 92

[28] Philippe Braunstein, 'Innovations in Mining and Metal Production in Europe in the Late Middle Ages', *Journal of European Economic History* 12 (1983), 581. On Nuremberg's relationships with Eastern Europe, see Ute M. Schwob, *Kulturelle Beziehungen zwischen Nürnberg und den Deutschen im Südösten in 14. bis 16. Jahrhundert,* Munich, 1969.

hand, while steadying the weapon with the other. The small *Hackenbüchse* lost its hook as it became more portable. Standard tables of shot weight from the Nuremberg inventory of 1462 mentioned above list *Hackenbüchse* bullets weighing from 75 gm (3 oz) down to 12.5 gm (½ oz).[29] This small 'hook gun' spread throughout Europe, along with its somewhat inelegant name. For the French, it became the *harquebuze*, while Dutch rendered it as *Hakkebuss*. English, naturally, called the gun by names derived either from the Dutch (*hackbutt* or *hagbutt*) or from the French (*arquebus*). A later shift in terminology would label all such shoulder arms simply 'muskets'.

The development of the arquebus or proto-musket in the fifteenth century provided yet another weapon that corresponded to the needs of the urban militias, and it is hardly surprising to find these guns being used by militia troops before they became commonplace in more professional mercenary armies. The association of arquebuses with city wall defense is manifest in the charmingly naïve drawings done for Benedicht Tschachtlan's *Berner Chronik* (ca. 1470). (fig. 10.1) In the final decades of the fifteenth century, the arquebus spread rapidly among militia troops. It was cheap to make, used no steel, only common iron, and cost less than half the price of a crossbow.[30] The Christian chronicles of the Granada War (1481-92) refer frequently to arquebusiers, a reflection of the local Spanish militia troops who joined in the final crusade against the Moors.[31] In 1490, the Venetian senate dispatched arquebus experts to all villages in Friuli to teach the new weapon to militia recruits. The muster of 1493 called for 1,000 handgunners (of a force of 4,000), and some 900 of them actually made an appearance.[32] Thirty years later, Luigi da Porto describes the inhabitants of Friuli in a letter as being skilled in the use of the arquebus both in hunting and in military

---

[29] For Conrad Gürtler's 1462 inventory, see Willers, *Nürnberger Handfeuerwaffe*, pp. 5 ff.

[30] In Spain, during the Granada War, the price for an *espringard* or *arcabuz* in 1485 was 341 maravedis or 11 reales. See Miguel Angel Ladero-Quesada, *Castilla y la conquista del Reino de Granada*, Valladolid, 1967, 128. Even elaborately finished presentation pieces, like the inlaid hunting arquebuses the Nuremberger patrician Anton Tucher sent the Bohemian noble Christoph von Schwanberg as a gift in 1519, cost only about 40 per cent of the price of similarly decorated crossbows (5 florins versus 13½ florins). See Willers, *Nürnberger Handfeuerwaffe*, pp. 17-8.

[31] Fernando del Pulgar, *Crónica de los Reyes Católicos*, v. 2, *Guerra de Granada*, Jan de Mata Carriazo, ed, v. 6, Madrid, 1943, 64, 224, and 288. See Weston F. Cook, Jr., 'The Cannon Conquest of Nasrid Spain and the End of the Reconquista', *Journal of Military History* 57 (1993) 43-70, esp. p. 46, n. 13.

[32] J.R. Hale, 'Brescia and the Venetian Militia System in the Cinquecento', in *Armi e Cultura nel Bresciano, 1420-1870*, Brescia, 1981, 97-119, see p. 98.

exercises.[33] By then, of course, arquebus troops were commonplace, and had proved their value in sieges and open battles alike.

CITIES AND FIREARMS: THE WHEEL-LOCK PISTOL

The city's particular combination of skills and markets made one final contribution to medieval and Renaissance warfare, the wheel-lock pistol. The matchlock involved a length of smoldering fuse held in a mechanism controlled by a trigger. The wheel-lock mechanism consists of a spring-driven steel wheel that struck sparks from a flint in a manner superficially resembling modern cigarette lighters.[34] The wheel lock was a precursor of the flintlock, but more important, it was the first mechanism that allowed a fully primed, ready-to-shoot gun that could be carried in a case or holster.

The earliest illustrations of a primitive form of wheel lock appear in a manuscript dateable to 1505 and associated with a Nuremberger patrician, Martin Löffelholtz.[35] This has fostered the thesis that the wheel lock was a German invention.[36] On the other side of the evidence is the appearance of wheel locks and associated mechanisms as well as indications of their fabrication in Leonardo da Vinci's *Notebooks* from about the same date.[37] For our purposes, this quarrel between Nuremberg and Milan makes no difference;

---

[33] Bartolomeo Bressan, ed, *Lettere storiche di Luigi da Porto*, Florence, 1857, Letter 44 (7 April 1510) 181.

[34] The resemblance is misleading in that a modern lighter's wheel is harder than its 'flint' (actually a soft cerium alloy), and it is this 'flint' that wears away. In a wheel lock, the pyrites are harder than the steeled wheel, and it is the wheel which gives off incandescent fragments of itself to make the sparks.

[35] Berlin, Staatsbibliothek Preussischer Kulturbesitz, Ms. germ quart 132, a 76-folio illustrated collection of miscellaneous drawings. The manuscript is untitled, but it has the heraldic arms of the Löffelholtz family and the date 1505 as well. The codex is described in Feldhaus, Franz Maria., 'Eine Nürnberger Bilderhandschrift', *Mitteilungen des Vereins für Geschichte der Stadt Nürnberg* 31 (1933) 222-26. The identification of Martin as the family member associated with the manuscript was made by Emil Reicke, 'Martin Löffelholtz, der Ritter und Techniker', *Mitteilungen des Vereins für Geschichte der Stadt Nürnberg* 31 (1933) 227-39. I erroneously identified this as Ms. germ. fol. in my *The Technological Illustrations of the So-Called 'Anonymous of the Hussite Wars'*, Wiesbaden 1979, 121. The manuscript itself was destroyed during World War II, but good-quality copies survived and microfilms made from these are quite legible.

[36] For a discussion of claims to priority, see Claude Blair, 'Further Notes on the Origins of the Wheellock', *Arms and Armor Annual*, 1 (1973) 28-47, and the further discussion by Vernard Foley, 'The Invention of the Wheellock', *Journal of the Arms and Armour Society* 11 (1985) 207-48.

[37] For evidence regarding Leonardo da Vinci's role in the invention of the wheel lock, see Vernard Foley et al., 'Leonardo, the Wheel Lock, and the Milling Process', *Technology and Culture* 24 (1983) 399-427.

the wheel lock was unequivocally an urban innovation. The technical skill needed to make a wheel lock was extremely high. Clearances between the wheel and its slot must be quite close, less than 0.04 mm (roughly 0.002"), or the mechanism will choke and jam.[38] It was an expensive device, one best made by sophisticated big-city gunsmiths. Of course, it was also in cities that the new gun spread, because it could be concealed in the clothing and used surreptitiously. The Augsburg chronicler Wilhelm Rem tells us the tale of one Laux Pfister, who took one on a trip to Constance in 1515. Unfortunately, Pfister was entertaining a 'handsome whore' in his room when he accidentally shot her through the jaw with his 'self-igniting gun'. She survived, but Pfister was fined and made to pay her an annuity in compensation. (He had, after all, ruined her livelihood.)[39] Incidents like these led to attempts to ban the new weapon, which was regarded as altogether too handy for highwaymen and robbers, but to no avail.

The wheel-lock pistol became the final step in a process of change that dispossessed the heavily armored medieval knight of the battlefield primacy he once enjoyed. By the mid-sixteenth century, light cavalrymen armed with pistols could match the mobility of the heavy cavalryman, but were his superior in firepower.[40] Nothing in the earlier spread of arms had quite tipped the balance as did these new guns, and by the late sixteenth century, a thousand years of chivalric warfare had given way to something rather different and certainly a great deal bloodier, thanks to the cleverness of urban gunsmiths and their infernal machines.

CONCLUSION

Cities, it seems, may have been undervalued as sources of technological innovation in warfare. They were not, to be sure, ever the source of the most fearsome troops or of overwhelming field armies, but they were hotbeds of technical innovation that fostered weapons suited to the specific needs, and the specific skills, of urban dwellers. Weapons requiring more skill to fabricate but less to operate became the principal types that were identified with cities

---

[38]   Foley et al., 'Leonardo', p. 403, indicates clearances in the range of 0.04-0.08 mm as against grain sizes tending down to about 0.1 mm for priming powders.
[39]   Blair, 'Further Notes', p. 35.
[40]   On the tactical significance of the wheel-lock pistol, see: Claude Gaier, 'L'opinion des chefs de guerre français du XVIe siècle sur les progrès de l'art militaire', *Revue Internationale d'Histoire Militaire* 29 (1970) 723-46; Claude Gaier, 'La cavalerie lourde en Europe occidentale du XIIe au XVIe siècle: Un problème de mentalité', *Revue Internationale d'Histoire Militaire* 31 (1971) 385-96; and Frederick J. Baumgartner, 'The Final Demise of the Medieval Knight in France', in Jerome Friedman, ed., *Regnum, Religio et Ratio: Essays Presented to Robert M. Kingdon*, Kirksville, MO, 1987, 9-17.

and their soldiers. This pattern was evident in the use of crossbows, later in arquebuses, and finally in the development of the wheel-lock pistol. Cities were often the focal point of elaborate and complex networks that brought together raw and semi-processed materials, skilled labor, and a certain informal 'managerial' ability to define what was needed. They also served to distribute the élite and commodity weapons that their workshops produced to a widespread clientele that seems to have been quite sophisticated about price-quality tradeoffs. In producing for this international market, and for local, domestic needs as well, cities like Nuremberg found themselves in the forefront of weapons development at a time when new weapons were helping to alter the oldest traditions of European warfare.

Index